Basic Geological Mapping

The Geological Field Guide Series

Basic Geological Mapping

FIFTH EDITION

Richard J. Lisle
School of Earth and Ocean Sciences, Cardiff University

Peter J. Brabham
School of Earth and Ocean Sciences, Cardiff University

and

John W. Barnes

WILEY-BLACKWELL

A John Wiley & Sons, Ltd., Publication

This edition first published 2011 © 2011 by John Wiley & Sons, Ltd.

Wiley-Blackwell is an imprint of John Wiley & Sons, formed by the merger of Wiley's global Scientific, Technical and Medical business with Blackwell Publishing.

Registered office: John Wiley & Sons, Ltd, The Atrium, Southern Gate, Chichester, West Sussex, PO19 8SQ, UK

Editorial Offices: 9600 Garsington Road, Oxford, OX4 2DQ, UK
The Atrium, Southern Gate, Chichester, West Sussex, PO19 8SQ, UK

111 River Street, Hoboken, NJ 07030-5774, USA

For details of our global editorial offices, for customer services and for information about how to apply for permission to reuse the copyright material in this book please see our website at www.wiley.com/wiley-blackwell.

The right of the author to be identified as the author of this work has been asserted in accordance with the UK Copyright, Designs and Patents Act 1988.

Library of Congress Cataloging-in-Publication Data

Lisle, Richard J.
  Basic geological mapping. – 5th ed. / Richard J. Lisle, Peter J. Brabham, and John W. Barnes.
    p. cm.
  Rev. ed. of: Basic geological mapping / John W. Barnes, with Richard J. Lisle. 4th ed.
  Includes index.
    ISBN 978-0-470-68634-8 (pbk.)
1. Geological mapping. I. Brabham, Peter. II. Barnes, J. W. (John Wykeham), 1921- III. Barnes, J. W. (John Wykeham), 1921- Basic geological mapping. IV. Title.
  QE36.B33 2011
  550.22′3−dc23

                                                                                2011022844

ISBN: 978-0-470-68634-8

A catalogue record for this book is available from the British Library.

This book is published in the following electronic format: ePDF 978-1-119-97402-4; ePub 978-1-119-97751-3; MOBI 978-1-119-97752-0

Typeset in 8.5/10.5pt Times by Laserwords Private Limited, Chennai, India.
Printed and bound in Singapore by Ho Printing Singapore Pte Ltd

4   2013

CONTENTS

CONTENTS

CONTENTS

CONTENTS

PREFACE TO THE FOURTH EDITION

This book is a basic guide to field techniques used in geological mapping. It is meant to be kept in camp with you and even carried in your rucksack in the field. In addition, because no piece of geological mapping can be considered complete until the geology has been interpreted and explained, chapters are provided on drawing cross-sections, on preparing and presenting 'fair copy' maps, and on presenting geological diagrams from your fieldwork suitable for inclusion in your report. A report explaining the geology is an essential part of any field project, and a brief chapter on the essentials for writing and illustrating it concludes this book. Some emphasis, too, is given to field sketch-mapping because many reports lack those large-scale detailed maps of small areas that can often explain complex aspects of the geology that cannot be shown on the scale of the field map being used, and that are difficult to describe in words. Attention is also given to field notebooks, which are, in many cases, deplorable.

It is assumed that readers of this book have already had at least one year of university or equivalent geology, and have already been told what to look for in the field. Geological mapping cannot, however, be taught in lectures and the laboratory: it must be learnt in the field. Unfortunately, only too often, trainee geologists are left largely to their own devices, to sink or swim, and to learn to map for themselves with a minimum of supervision on 'independent' mapping projects. It is hoped that this book will help in that task.

John W. Barnes and Richard J. Lisle
2003

PREFACE TO THE FIFTH EDITION

This fifth edition of *Basic Geological Mapping* was revised without the help of John Wakeham Barnes who sadly passed away in 2007. On the suggestion of referees we have updated the text by including mention of modern technological aids and data that are used in modern geological mapping, such as applied geophysics, digital terrain models, and optical and GPS-based surveying. There has also been more emphasis given to structural geology and cross-section construction. Whilst making these additions we have been conscious of the need to maintain John's successful formula; to offer a practical guide for the student undertaking a mapping project with a minimum of resources and academic supervision.

We are grateful for the advice of Jim Hendry and Rob Strachan (University of Portsmouth), Chris Berry and Alun Rodgers (Cardiff University) and the book's reviewers.

Richard J. Lisle and Peter J. Brabham
January 2011

1

INTRODUCTION

Most geological maps record the regional distribution of rocks belonging to different formations. However, such maps reveal far more than where we could find rocks belonging to a given formation. The geometrical shape of the different formations on the geological map can also be interpreted in terms of the geological structure and geological history of the region concerned. As an earth scientist you must remember that *accurate* geological maps form the basis of most geological work, even laboratory work. They are used to solve problems in earth resource exploration (minerals and hydrocarbons), civil engineering (roads, dams, tunnels, etc.), environmental geoscience (pollution, landfill) and hazards (landslides, earthquakes, etc.). Making a geological map is therefore a fundamental skill for any professional geologist. As Wallace (1975) states, 'There is no substitute for the geological map and section – absolutely none. There never was and there never will be. The basic geology still must come first – and if it is wrong, everything that follows will probably be wrong.'

There are many kinds of geological map, from small-scale reconnaissance surveys to large-scale detailed underground maps and engineering site plans, and each is made using different techniques. In this textbook, however, we are concerned only with the rudiments of geological mapping. The intention is to provide basic methods and good field practice on which you can further build, and adapt, to deal with a wide range of types of geological mapping.

1.1 Outline and Approach

This book is arranged in what is hoped is a logical order for those about to go into the field on their first independent mapping project. This first chapter includes the important issue of fieldwork safety and appropriate conduct during fieldwork, which should always be considered before anything else. The equipment you will need for mapping is described in Chapter 2, which is followed by a chapter devoted to the many types of geological map you may have to deal with some time during your professional career. A description follows of the different kinds of topographic base maps that may be available on which to plot your geological observations in the field. Methods to locate yourself on a map are

Basic Geological Mapping, Fifth Edition.
Richard J. Lisle, Peter J. Brabham and John W. Barnes.
© 2011 John Wiley & Sons, Ltd. Published 2011 by John Wiley & Sons, Ltd.

also described, and advice is given on what to do if no topopgraphic base maps at all are available.

The following four chapters describe the methods, techniques and strategies used in geological mapping, including a brief description of photogeology – that is, the use of aerial photographs in interpreting geology on the ground. A further chapter is devoted to the use of field maps and those most neglected items, field notebooks.

The last three chapters concern 'office work', some of which may have to be done whilst still at your field camp. They cover methods of drawing cross-sections and the preparation of other diagrams to help your geological interpretation. Advice is also given on preparing a 'fair copy' geological map that shows your interpretation of the data from your field map. However, a geological map is not, as is sometimes supposed, an end in itself. The whole purpose is to explain the geology of the area and your map is only a part of that process: a report is also needed to explain the geological phenomena found in the area and the sequence of geological events. Chapter 11 is a guide on how to present this important part of the geological mapping project.

The approach here is practical: it is basically a 'how to do it' book. It avoids theoretical considerations. It is a guide to what to do in the field to collect the evidence from which conclusions can be drawn. What those conclusions are is up to you, but bear in mind what the eminent geologist Lord Oxburgh has said about mapping – that making a geological map is one of the most intellectually challenging tasks in academia (Dixon, 1999).

1.2 Safety
DO NOT PROCEED UNTIL YOU HAVE READ THIS SECTION!

Geological fieldwork is not without its hazards. In Britain, field safety is covered by the Health and Safety at Work Act 1974, and its subsequent amendments. Both employers and workers have obligations under the Act and they extend equally to teachers and students.

The safety risks depend on the nature of the fieldwork as well as on the remoteness, weather conditions and topography of the area being mapped. Before starting the mapping project, a formal *risk assessment* should be carried out. This will determine the safety precautions and the equipment to be carried whilst in the field. Table 1.1 lists some common risks, but your risk assessment must also consider the specific dangers associated with the area to be mapped. This will involve doing your homework before leaving for the mapping, for example consulting topographic maps, finding the address of nearest medical services, looking at tide tables, and so on.

Table 1.1 *Common safety hazards associated with geological mapping.*

Risk	Precautions
Fall from steep slopes	Stay away from cliffs, steep slopes, quarry edges, overgrown boulder fields, and so on. Do not rely on Global Positioning System (GPS) but examine a topographic map to identify steep slopes and plan your route. Avoid climbing; leave dangerous exposures unmapped rather than take risks. Do not run down slopes. In mountains but not on a path, stay put in dense mist, fog and darkness
Struck by falling rock and splinters from hammering	Avoid rock overhangs; wear a helmet if near cliffs, quarry faces. Do not enter mines or caves. When hammering always use safety goggles and take care with bystanders and passers-by
Drowning after being swept away by waves, tides and floods	Avoid the water's edge at sea, lakes and rivers. Consult tide tables. Do not enter caves, mines, potholes. Do not attempt to cross fast-flowing rivers
Cannot be reached by emergency services	Work in pairs, or in close association; leave details of the day's route in camp before leaving for the field; wear bright clothing, carry a mobile phone, whistle, torch, flashing LED beacon or a mirror to attract the attention of passers-by or mountain rescue teams
Exposure, an extreme chilling arising from sudden drop in temperature	The symptoms range from uncontrolled shivering, low body temperature, exhaustion and confusion. Carry warm clothing and waterproofs, thermal safety blanket, matches, emergency rations (e.g. glucose tablets, water)
Motoring accident	Drive carefully on narrow mountain roads; at roadside exposures take care with passing traffic and wear high-visibility jackets. Never drive whilst under the influence of alcohol or drugs.

A geologist should be able to swim, even if fully clothed. If you swim you are less likely to panic when you slip off an outcrop into a river; or from weed-covered rocks into the sea or a rock pool, or even if you just fall flat on your face when crossing a seemingly shallow stream. Such accidents happen to most of us sometime. If you are faced by something risky, play it safe, especially if you are on your own. A simple stumble and a broken ankle in a remote area can suddenly become very serious if nobody knows where you are and you are out of mobile phone coverage.

In some northern latitudes (e.g. northern Canada, Svalbard) geologists have to carry guns and flares to ward off the unwanted attentions of polar bears. So if you are planning work abroad, do your homework on special dangers before you go.

1.3 Field Behaviour

Geologists spend much of their time in the open air and, more often than not, their work takes them to the less inhabited parts of a country. If they did not like being in open country, presumably they would not have become geologists in the first place: consequently, it is taken for granted that geologists are conservation-minded and have a sympathetic regard for the countryside and those who live in it. Therefore, remember the following:

1. Do not leave gates open, climb wire fences or drystone walls or trample crops, and do not leave litter or disturb communities of plants and animals.
2. Do not hammer for the sake of it. Greenly and Williams (1930, p. 289) observe that 'indiscriminate hammering is the mark of a beginner' (several key localities once showing beautiful structures have been defaced by geological hammering, drilling and graffiti). When you are collecting specimens do not strip out or spoil sites where type fossils or rare minerals occur. Take only what you need for your further research.
3. Before you embark on any field programme you should have studied your public access rights on footpaths using maps or web-based enquiry. In the UK, you do not have the right to walk wherever you want, but open access to many remote areas is now covered by the Countryside and Rights of Way Act 2000. These are typically areas of mountain, moor, heathland, downland and registered common land; further details can be found on the Ramblers web-site. When in the field always ask permission to enter any private land when not on a public footpath. Most owners are willing to cooperate with geology students if they are asked politely first; landowners are usually very interested in what lies beneath their land, but understandably get very annoyed to find strangers sampling their rocks uninvited.

If working in a foreign country, carry a simple A5 size laminated card explaining in the local language who you are and what you are doing; this often diffuses any conflict and confusion with landowners due to your poor communication skills. Bear in mind that irate farmers can inhibit/restrict geological activities in an area for years to come, and this has already happened in parts of Britain. Many other countries are less populated and have open space, and the situation may be easier, but every country has some land where owners expect you to consult them before working there. If in doubt, ask! (See also the *Geological Fieldwork Code* published by the Geologists' Association, 2000.)

1.4   A Few Words of Comfort

Finally, some cheering words for those about to start their first piece of independent mapping. The first week or so of nearly every geological mapping project can be depressing, especially when you are on your own in a remote area. No matter how many hours are spent in the field each day, little seems to show on the map except unconnected fragments of information that have no semblance to an embryonic geological map. Do not lose heart: this is quite normal. Like solving a jigsaw, the first stages are always slow until a pattern starts to emerge; then the rate of progress increases as the separate pieces of information start fitting together.

The last few days of fieldwork are often frustrating for, no matter what you do, there always seems to be something left to be filled in. When this happens, check that you do have all the essential information and then work to a specific finishing date. Otherwise you will never finish your map.

Detailed fieldwork preplanning, executing a daily field plan and good time management are often the keys to success.

2

FIELD EQUIPMENT

Geologists need a number of items for the field. A hammer (sometimes two) is essential and some chisels. Also essential are a compass, clinometer, pocket steel tape and a hand lens, plus a map case, notebooks, map scales, protractor, pencils and eraser, an acid bottle and a penknife. A camera is a must, and a small pair of binoculars can be useful at times, for studying dangerous or inaccessible cliffs from a distance.

A GPS is very useful for field mapping, but it must never be totally relied upon as your only way of determining where you are. Remember GPS units do not work if their batteries run down or if there is minimal sky cover, that is in forests, quarries, and so on (see Section 3.5.10).

A 30 m tape will sometimes be needed and a stereographic net. If using aerial photographs you may need a pocket stereoscope or a pair of stereo glasses. You will also need a felt-tipped marker pen for labelling specimens. Make sure it writes on plastic and is totally waterproof. Finally, you will need a rucksack to carry it all, plus a water bottle, emergency rations, a first-aid kit, whistle, perhaps an extra sweater, your mobile phone and, of course, your lunch.

Geologists must also wear appropriate clothing and footwear for the field if they are to work efficiently, often in wet cold weather when other (perhaps more sensible) people stay indoors; inadequate clothing can put a geologist at risk of hypothermia (see Section 1.2). A checklist of what you may have to pack before a field trip is given in Appendix B, but this is an exhaustive list to cover various types of geological fieldwork in various climates; refer to it before setting out to your field area base.

2.1 Hammers and Chisels

Any geologist going into the field needs at least one hammer with which to break rock. Generally, a hammer weighing less than about 0.75 kg (1.5 lb) is of little use except for very soft rocks; 1 kg (2–2.5 lb) is probably the most useful weight. The commonest pattern still used in Europe has one square-faced end and one chisel end. Many geologists now prefer a 'prospecting pick'; it has a long pick-like end that can be inserted into cracks for levering out loose rock, and can also be used for digging though a thin soil cover. Hammers can be

Basic Geological Mapping, Fifth Edition.
Richard J. Lisle, Peter J. Brabham and John W. Barnes.
© 2011 John Wiley & Sons, Ltd. Published 2011 by John Wiley & Sons, Ltd.

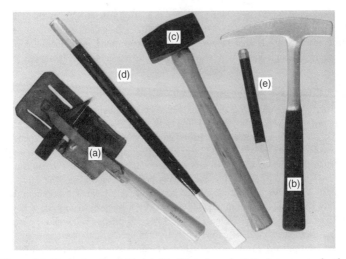

Figure 2.1 *Tools for the field. (a) Traditional geologist's hammer in leather belt 'frog'; (b) steel-shafted 'prospecting pick'. (c) Bricklayer's 'club' hammer with a replaced longer shaft. (d) A 45 cm chisel with 2.5 cm cutting edge and (e) An 18 cm chisel with 2 cm edge. (After Cooper, G.R. and C.D. McGillem, 1967:* Methods of Signal and System Analysis. Holt, Rinehart and Winston, New York, 432 pp.)

bought with either wood or fibreglass handles, or with a steel shaft encased in a rubber grip (Figure 2.1).

Geologists mapping hard igneous and metamorphic rocks may opt for heavier hammers. Although 2 kg (4 lb) geological hammers are available, a bricklayers 'club' hammer, with a head shaped like a small sledgehammer, can be bought more cheaply; but replace its rather short handle by a longer one bought from a hardware store.

Hammering alone is not always the best way to collect rock or fossil specimens. Sometimes a cold chisel is needed to break out a specific piece of rock or fossil. The size of chisel depends on the work to be done. Use a 5 mm (1/4 inch) chisel to delicately chip a small fossil tree from shales; but to break out large pieces of harder rock a 20–25 mm (3/4 inch) chisel is required (Figure 2.1). One thing you must never do is to use one hammer as a chisel and hit it with another. The tempering of a hammer face is quite different from that of a chisel head, and small steel fragments may fly off the hammer face with unpleasant results. Eye damage due to rock or metal splinters is often permanent, so again always wear safety goggles when hammering.

Some geologists carry their hammers in a 'frog', or holster, as this leaves their hands free for climbing, writing and plotting. They can be bought or easily made from heavy leather (Figure 2.1). Climbing shops stock them for piton hammers although some may be too small to take a geological hammer.

2.2 Compasses and Clinometers

The perfect geologist's compass has yet to be designed. Americans have their Brunton, the French the Chaix-Universelle, the Swiss have their Meridian, whilst the Germans have their Breithaupt Clar compass. All are expensive. Many geologists now use the very much cheaper mirror compasses: the Swedish Silva 15TD/CL, the Swiss Recta DS50 or the similar Finnish Suunto MC2 (Figure 2.2a). All of the above have built-in clinometers. The Silva and Suunto compasses (Figures 2.2 and 2.3), however, have a transparent base so that bearings can be plotted directly onto a map by using the compass itself as a protractor (see Figure 3.4). However, prismatic compasses, which have a graduated card to carry the magnetic needle, are perhaps easier for taking bearings on distant points. All these compasses except the Brunton are liquid-filled to damp the movement of the needle when taking a reading. The

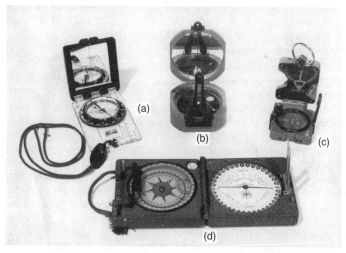

Figure 2.2 *Compasses designed for the geologist. (a) Finnish Suunto compass, similar to the Swedish Silva Ranger 15 TDCL. (b) American Brunton 'pocket transit'. (c) Swiss Meridian compass and (d) French Chaix-Universelle. The Brunton and Meridian can also be used as hand levels.*

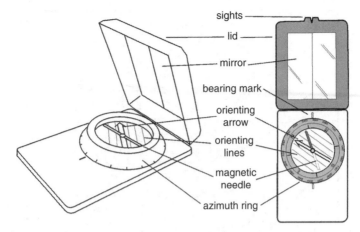

Figure 2.3 *The mirror compass. (After Cooper, G.R. and C.D. McGillem, 1967:* Methods of Signal and System Analysis. *Holt, Rinehart and Winston, New York, 432 pp.)*

Brunton is induction-damped. Some compasses can be adjusted for work in variable latitudes (Recta, Silva).

2.2.1 Compass graduations

Compasses can be graduated in several ways. The basic choice is between the traditional degrees and continental grads. There are 360° in a full circle, but 400 grads. Both are used in continental Europe and if you do buy a compass in Europe, check it first. If you opt for degrees, you must then choose between graduation in four quadrants of 0–90° each or to read a full circle of 0–360° (azimuth graduation). We recommend using the azimuth, since bearings can be expressed more briefly and with less chance of error and confusion. Comparisons are made in Table 2.1.

Table 2.1 *Equivalent bearings using quadrant and azimuth conventions.*

Quadrant bearing	Azimuth bearing
N36°E	036°
N36°W	324°
S36°E	144°
S36°W	216°

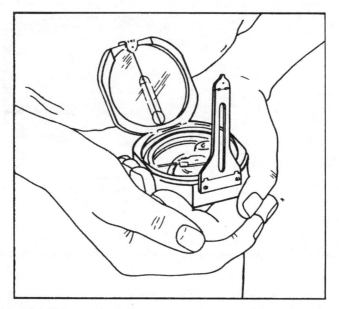

Figure 2.4 *The recommended way to use a Brunton compass when taking a bearing on a distant point.* (Reproduced by courtesy of the Brunton Company, Riverton, Wyoming.)

2.2.2 Using compasses

Prismatic compasses and mirror compasses are used in different ways when sighting on a distant point. A prismatic is held at eye level and aimed like a rifle, lining up the point, the hairline at the front of the compass and the slit just above the prism. The bearing can be read in two ways. The Brunton Company recommends that the compass is held at waist height and the distant point aligned with the front sight so that both are reflected in the mirror and are bisected by the hairline on the mirror (Figure 2.4).

With the Silva-type mirror compass, sight on the distant point by holding the compass at eye level and reflecting the compass needle in the mirror (Figure 2.5). In fact, some prefer to read a Brunton in the same way. Mirror compasses have a distinct advantage over prismatic compasses in poor light such as underground in mines. Specialist mining compasses can be used in the dark by pressing a button to clamp the compass reading, so you can then study the compass needle under the light of a cap lamp.

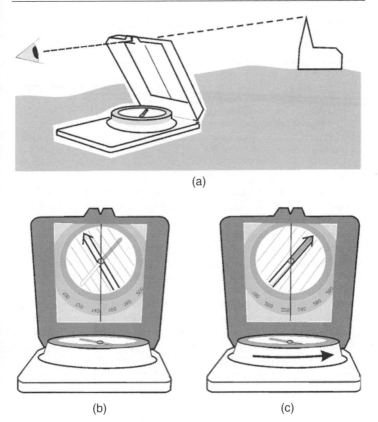

Figure 2.5 *Taking a bearing with a mirror compass: (a) sighting on a landmark; (b) view through the mirror and (c) rotating the azimuth ring until the orienting arrow aligns with the magnetic needle.*

Taking a bearing with a mirror compass

1. Hold the compass level in an outstretched hand (Figure 2.5a). This will mean first removing the chord from your neck.
2. Tilt the lid of the compass so that the compass dial is fully visible through the mirror (Figure 2.5b).

3. Aim the compass at the landmark using the sights on the lid.
4. A correct alignment of the compass is achieved by looking in the mirror, and by rotating the whole compass in your hand until the line drawn on the mirror appears to pass through the central point on the magnetic needle (Figure 2.5b).
5. Maintaining the correct aim and alignment of the compass, rotate the azimuth ring (Figure 2.5c) until the orienting arrow aligns with the magnetic needle with the arrow head at the north end of the needle.
6. The bearing can be read off the azimuth ring at the bearing mark.

2.2.3 Clinometers

Clinometers are used to measure the angle of dip of a planar structure or the plunge angle of a linear structure. If your compass does not incorporate a clinometer, the latter can be bought separately. A variety of cheap clinometers are on the market, the majority designed for use by builders and engineers. They are sold under various names: angle indicators, inclinometers, digital spirit levels and digital angle gauges. A digital clinometer can now also be obtained as a mobile phone app, but these can be unreliable without calibration.

Clinometers can be easily made, either by using the pendulum principle (Figure 2.6a) or, better still, the Dr Dollar design (Figure 2.6b), as follows: photocopy a 10 cm diameter half-round protractor for a scale and glue it to a piece of Perspex after removing the figures and renumbering so that 0° is now at the centre. Cement transparent plastic tubing containing a ball-bearing around it and fill each end with plasticine or putty to keep the ball in (Barnes, 1985). Alternatively, Dr Dollar clinometers with the name Maxiclin can be bought.

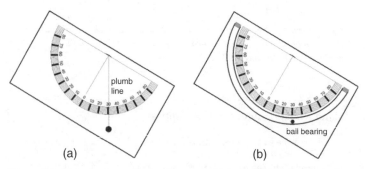

Figure 2.6 Ideas for home-made clinometers: (a) plumb-line type and (b) ball bearing in transparent plastic tube.

2.2.4   Lineation compass

The Japanese produce a most useful compass designed to measure the trend and plunge of a lineation simultaneously. The case of this 'universal compass' is on gimbals so that it always remains level whatever the angle of its frame (Figure 2.7). It is effective in even the most awkward places. The design is derived from Ingerson (1942). The maker is Nihon Chikagasko Shaco, Kyoto. Geoclino, a digital version of the universal compass, is made by GSI of Japan.

2.3   Hand Lenses

Every geologist must have a hand lens and should develop the habit of carrying it at all times (Figure 2.8). A magnification of between 7 and 10 times is probably the most useful. Although there are cheap magnifiers on the market, the flatness of field obtained with a good-quality lens is worth the extra cost, and such a lens should last you a lifetime. To ensure that it does last a lifetime, attach a thin cord to hang it around your neck. Monocle cord is ideal if you can find it, as it does not twist into irritating knots. However, always keep a spare back at

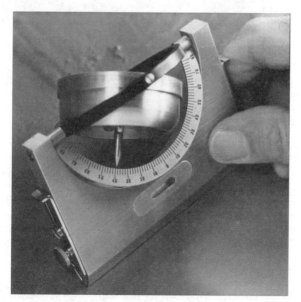

Figure 2.7 *Japanese 'universal clinometer'. Plunge direction can be measured directly from the compass, which always stays horizontal, and plunge is read by the pointer hanging below the compass box.*

Figure 2.8 *The hand lens is an essential tool for close-up study of rocks in the field.*

basecamp – your fieldwork could be jeopardised should you lose the only one you have with you.

2.4 Tapes

A short retractable steel tape has many uses. A 3 m tape takes up no more room than a 1 m tape and is much more useful. You can use it to measure everything from grain size to bed thickness, and if the tape has black numbering on a white background, you can use it as a scale when taking close-up photos of rock surfaces or fossils. A geologist also occasionally needs a 10 m or 30 m fibreglass tape for small surveys. You might not need it every day, but keep one back in basecamp just in case.

2.5 Map Cases

A key necessity for geological mapping is a map case of appropriate design. A map case is obviously essential where work may have to done in the rain or mist; but even in warmer climes, protection from both the sun and sweaty hands is still needed. However, some map cases made for hikers are basically just a

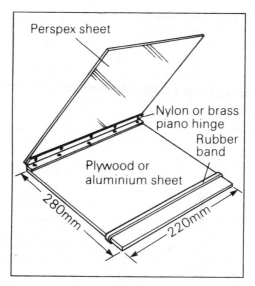

Figure 2.9 A map case made from a Perspex sheet attached to a plywood base by a nylon or brass 'piano hinge' (DIY shop) and 'pop' rivets.

transparent plastic map holder; they are of little use. A geologist's map case must have a rigid base so that you can plot and write on the map easily. It should have a transparent cover that allows you to see the map without exposing it to the elements and sweaty hands. This see-through lid must open easily, otherwise it will deter you from adding information to the map. If it is awkward to open, you will probably say 'I will remember that and add it later' and of course, being only human, you forget! The best map cases are probably home-made (Figure 2.9).

Pencil holders make mapping easier, whether attached to your map case, on your belt or as part of your mapping jacket. Continually groping around in anorak pockets for a particular coloured pencil can be very wasteful in fieldwork time.

2.6   Field Notebooks

Never economise on your field notebook. It should have good-quality 'rainproof paper', a strong, hard cover and good binding. It will have to put up with hard usage often in wet and windy conditions. Nothing is more discouraging than to see pages of your field notes torn out of your notebook by a gust of wind and blowing across the hillside. Loose-leaf books with spiral binding are especially vulnerable. A hard cover is necessary to give a good surface for writing and sketching. Ideally a notebook should fit into your anorak pocket so that it is always available, but big enough to write on in you hand. A good size is 12×20 cm, so make sure you have a pocket or belt-pouch to fit it. Try to buy a book with squared, preferably metric squared, paper; it makes sketching so much easier. Half-centimetre squares are quite small enough. A surveyor's chaining book is the next best choice: the paper is rainproof, it is a convenient size and it has a good hard cover. A wide elastic band will keep the pages flat and also mark your place so that you are not continually looking for the current notebook page.

2.7 Scales

A geologist must use suitable scales, most conveniently about 15 cm long; a long ruler is just not good enough. Rulers seldom have an edge thin enough for accurate plotting of distances, and trying to convert in your head a distance measured on the ground to the correct number of millimetres on the ruler for the scale of your map just leads to errors. Scales are not expensive for the amount of use they get. Many are thinly oval in cross-section and engraved on both sides to give four different graduations. The most convenient combination is probably 1:50 000, 1:25 000, 1:12 500, 1:10 000. In the USA, scales with 1:62 500 and 1:24 000 are needed. Colour code the edges by painting each with a different coloured waterproof ink or coloured adhesive tape – even nail varnish can be used – so that the scale you are currently using is instantly recognisable. Although triangular scales with six edges, each with a different scale, may seem an even better bet and are excellent for the drawing office, experience has shown that their knife-sharp edges are easily chipped in the field (Figure 2.10a).

The American transparent scale/protractor shown in Figure 2.10c only has two scales on it, but they are available with many different combinations of scales and are cheap enough so that a selection with different graduations can be bought.

2.8 Protractors

Little needs to be said about protractors. For ease of plotting they should be 12–20 cm in diameter and semicircular; circular protractors are no use for plotting in the field (see Section 6.2). Keep a couple of 10 cm protractors in your field kit in case of loss. Transparent protractors (and scales) are difficult to see when dropped in the field but are easier to find if marked with an orange fluorescent spot. If you do lose your protractor, the mirror compass can be used instead (Figure 2.11). Align the long edge of the compass with one line and then turn the azimuth dial until the orientation lines become parallel to the second

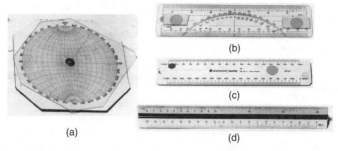

Figure 2.10 Pocket stereonet and scales (a) Home-made stereonet for the field; the upper rotating Perspex disc is slightly sandpapered so that it can be drawn on and easily cleaned off again. (b) Transparent combination of map scale and protractor (C-Thru Ruler Company). (c) Plastic scale with different graduations on both edges and both sides. (d) Triangular map scale, which is not recommended for field use but is excellent in the office.

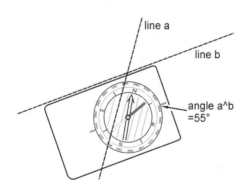

Figure 2.11 Using a mirror-type compass as a protractor.

line. The angle between the two lines is then given on the azimuth ring at the bearing mark.

2.9 Pencils, Erasers and Mapping Pens

At least three good-quality graphite pencils are needed in the field for mapping: a hard pencil (2H or 4H) for plotting bearings; a softer pencil (H or 2H) for plotting strikes and writing notes on the map, and another pencil (2H, HB or F) kept only for writing notes in your notebook. The harder alternatives are for warmer climates, the softer for cold. Do not be tempted into using soft (B) pencils; they smudge and need frequent sharpening. A soft pencil is quite incapable of making the fineness of line needed on a geological map with sufficient permanency to last a full day's mapping in rigorous conditions. Also, keep a separate pencil for your notebook to avoid frequent sharpening. Buy only good-quality pencils, and if possible buy them with an eraser attached; alternatively buy erasers that fit over the end of the pencil. Attach a larger, good-quality eraser to your buttonhole or your map case with a piece of string or cord, and always carry a spare. Coloured pencils should also be of top quality; keep a list of the make and shade numbers you do use so that you can replace them with exactly the same shades. In case of loss during mapping, it may be a good idea to have an identical set of coloured pencils back in camp.

You will also need mapping (or technical) pens for drawing lines of different thicknesses. In the past, geologists used refillable ink pens (e.g. Rotring), but these are now very expensive and difficult to obtain. There are many types of cheaper, disposable types available from good art and design shops (e.g. Staedtler pigment liner, Pilot drawing pens). The Edding Profipen 1800 comes in a range of colours of ink. Do not use technical pens in the field. They may be capable of fine lines and printing on a dry map, but not if the map is damp. Also Murphy's Law ensures that some notes are written just where a critical exposure will be found later in the day and the notes must be erased and rewritten elsewhere. Think carefully before you ink in your map as it is impossible to erase these permanent inks without damaging the surface of the map.

If you are collecting specimens in plastic sample bags you will need a thick permanent black marker pen that writes onto the plastic bag.

Tip: Before leaving for the fieldwork, try out your drawing pens; are they waterproof? Do they give crisp clean lines on the paper type to be used?

2.10   Acid Bottles

Always carry a small acid bottle in your rucksack. The bottle should contain a small amount of 10% hydrochloric acid and should be labelled as such. Five millilitres (5 mL) is usually ample for a full day's work even in limestone country, providing only a drop is used at a time – and one drop should be enough. Those tiny plastic dropping bottles in which proprietary eardrops and eyedrops are supplied make excellent field acid bottles. Make sure, however, that you change the label to show it now contains 10% acid. These bottles have the advantage that they deliver only one drop at a time, are small, do not leak and will not break. Keep a supply of 10% acid back in camp in a secure labelled childproof bottle.

2.11 Global Positioning System (GPS) and Mobile Phones

Today city-dwellers take it for granted that they can use their mobile phones in all locations (except on underground trains) and at all times. We have become totally reliant on them for urban life. When working as a geologist out in the countryside there is absolutely no guarantee your mobile phone will work. Even if it does, coverage will vary between network operators and be locally patchy. For example, you may obtain coverage whilst on top of a hill but not in a valley. Even in the UK, there are often many blind spots when mapping coastal sections. You can get some general idea of a particular network operator's coverage for your mapping area by looking at maps found on their website. In very remote areas only satellite phones will work and these are very expensive. Inexpensive two-way radios can be purchased for as little as £20, and these can be useful and cheaper to use whilst mapping in pairs or carrying out surveying exercises.

Do not presume in advance that your mobile phones will work in your mapping area, and you must never rely on them as your only safety device. If mapping in pairs do not assume you can just ring up your partner if you are separated in the field. Finally, make sure you have means of charging your mobile phone every evening. If camping out in very remote areas you may consider purchasing solar-panel battery-charging systems, which can be purchased from around £10 upwards. Spare batteries can be charged during the day when you are out working.

At times, locating yourself on a map can be time-consuming, especially where the base map lacks detail such as on open moorland or in deserts. For this reason geologists are increasingly making use of GPS to locate themselves in the field. GPS is a multi-satellite-based radio-navigation, timing and positioning system. It allows a person with a ground receiver to locate themselves anywhere on Earth in three dimensions (latitude, longitude and height above a global datum WGS-84) night or day. GPS was developed by the US government for military use, but since 1995 has also become available for civilian use. Initially its precision was deliberately degraded for civilian use, but since year 2000 civilians have been allowed greater functionality (but still not the full military accuracy, which is code protected). GPS is not the only available satellite navigation system; there are the Russian GLONASS and the European Union's GALILEO project (which is still under development). However, the cheapest navigation devices for field mapping are basic hand-held GPS units; these are called autonomous systems. Therefore we will concentrate on understanding the limitations of this system. Details of some of the more advanced (and more precise) GPS systems can be found in Chapter 5 of this book.

Dedicated field-rugged GPS systems used to be very expensive, but can now be purchased new from £50 to £250 (Figure 2.12). Many use simple rechargeable or disposable AA battery systems. In remote areas, maintaining battery power is

Figure 2.12 Some of the wide variety of hand-held GPS systems available on the market. (a) Basic rugged 12-channel Garmin GPS, which only provides locational coordinates but can be used with pocket PC (d). (b) Garmin GPS unit with ability to plot location onto a simple on-screen base map. (c) Mobile phone with GPS app installed. (d) Field-rugged Trimble pocket PC, which can take input from a GPS via cable or Bluetooth connection. A georeferenced air photo is shown on the screen. The software installed on the PC has the ability to log positional information along with field notes. Data can then be later automatically integrated into GIS software applications.

the biggest problem with any GPS unit, and you should have recharging facilities at basecamp (mains electricity or solar panels) or plenty of spare disposable battery sets. In subzero temperatures keep your GPS warm within your body clothing as battery efficiency reduces greatly in extreme cold.

Inexpensive GPS units will give you just your positional coordinates, whilst the more expensive units can plot your current position onto preloaded base maps or air photographs viewed on a small screen. All units have the capability of storing a large number of locational waypoints in a memory. When out mapping, your paper base map will probably use a local map grid system (e.g. UK Ordnance Survey), but the GPS fundamentally provides global latitude and longitude locations. It is vital that the GPS latitude/longitude output can be recalculated by the handset into the exact local map coordinate system that you are using on your base map. If not, you will lose most of the useful mapping functionality of the GPS; it will become a purely safety device. The library of available map conversions within each unit differs, so check your particular GPS unit's suitability for the country you are planning to work in. Useful technical GPS reviews, resources and practical information on using GPS with various countries' map coordinate systems can be found by web searching 'geocaching', which is a popular GPS treasure-hunting hobby. Manufacturers such as Garmin produce an ever changing range of field-rugged GPS systems with different levels of specification, so do your technical research before purchasing a GPS unit. GPS watches are also available but are more targeted at the long-distance running market.

Many GPS apps can also be found for iPhones and Android mobile phones but they should always be tested first for their reliability and accuracy. Many GPS apps have no map coordinate conversion capabilities. Mobile phones not only use the satellite GPS signals, but also know roughly where they are relative to the local mobile phone transmitters. This allows the mobile phone GPS to be more accurate in cities with tall buildings and a limited skyview, but this assisted Global Positioning System (A-GPS) function is much less useful in remote areas where transmitters are many miles apart or even non-existent. Mobile phones are not designed to be dropped onto rocks, so using your mobile continually as a GPS in the field is not good practice and is also a rapid battery drain. Use your mobile phone only for emergencies and to preserve battery power for when you really need it. Also, experience has shown that having fancy colour GPS app graphics on a mobile phone screen does not necessarily mean that the information projected is correct.

Your hand-held autonomous GPS receives signals from 24 satellites using six different orbits of the Earth (four satellites in each orbit plane). Their mean altitude is 20 200 km above the Earth with an orbital period of 12 hours. Each satellite contains an atomic clock and broadcasts a continual stream of signals. Your hand-held unit locates itself relative to the satellites by constantly measuring its distance (using the time delay of the signal from the satellite to the ground receiver) to as many moving satellites as possible. The satellites are constantly changing their configuration above the GPS handset on the Earth's surface and this can be plotted onscreen by the GPS. At least four satellites need to be visible in the sky to get a positional fix, and six visible satellites are recommended for an accurate fix. If a GPS unit cannot communicate with at least four satellites it will not give a positional fix. As these satellites are constantly in orbit in the sky and at any one time 12 of the 24 are likely to be out of sight around the other side of the Earth, to gain an accurate fix you need as wide as possible sky visibility. Also the longer you maintain a fix, generally the more precise your position becomes. Therefore if your sky view is limited because you are working in a deep ravine, under a cliff or in a forest and you cannot fix onto at least four satellites you will not get a positional fix, or at best a very imprecise location. In this case the geologist on the ground has to move to a better position such as the top of a hill, or forest clearing to get better sky visibility.

It is important to remember that although your basic autonomous GPS will give you an accurate location apparently to a within 1 m in latitude and longitude, the actual precision of that location will be much less. When you are locked onto six or more satellites for a few minutes, initially the GPS location precision will be around 150 m, precision will gradually improve to be on average around 10 m, and the best you can ever hope for is 5 m. The precision of the z dimension (height above a global datum) is typically twice as imprecise using a basic autonomous hand-held unit (around 15–20 m) and should not be relied on for any serious field surveying. It is absolutely essential that field geologists understand the variable precision of their GPS. The precision of any location is expressed as dilution of precision (DOP) or is more easily translated by the GPS into plotting a point on a base map surrounded by a circle with the radius of precision. Remember you could actually be anywhere within that circle of precision. You can test this variable precision easily using a GPS with a base map on a screen (or a paper map) and walk about and observe where your GPS actually plots your location whilst you are standing at a known point (e.g. a road intersection or trig point). You can also note that if you stay absolutely still, your GPS coordinates will change with time, oscillating around a point.

With good sky coverage a basic autonomous hand-held GPS will locate a geologist in the field to a latitude/longitude precision of on average around 10 m (about the size of a tennis court). Given this level of precision it cannot be used as an electronic tape measure, or to create very accurate topographic cross-sections. However, there are a number of very useful functions a GPS can perform to improve the efficiency and safety of field mapping:

- Always enter the location of your basecamp in the GPS memory, so if you get totally lost, at least you can use the GPS to walk in the right direction back to camp.
- If you walk continually with the GPS in your hand it will track its position and it can be used in electronic compass mode. This is useful for walking rapidly over monotonous open moorland along a set bearing, especially in disorientating misty conditions.

- You can simply log the grid references of all the outcrops that you have made notes of in your notebook by storing waypoints on the GPS and then also noting the waypoint number in your notebook. It is good practice to write these locations off the GPS into a notebook back at basecamp every evening, just in case you lose the GPS or erase its memory by accident.
- You can log the locations of specimen finds (fossils, minerals, etc.) as waypoints so that you can easily find these locations at a later date to collect more samples.
- If you wish to find an outcrop that you have seen on Google Earth imagery you can enter in the grid reference of that outcrop into the GPS and then use the GPS in the field to walk rapidly to that outcrop. Most GPS units will show a directional arrow pointing to the pre-entered ground location whilst you are walking with the GPS held out in front of you.

GPS units are not absolutely essential for field mapping in the UK, but are obviously essential if you are working in very remote locations such as deserts. For mapping in the UK, GPS units can be a very useful extra safety aid and improve your mapping efficiency. You should, however, never totally rely on a GPS for knowing where you are in the field. At any time your GPS battery could expire and you must then resort to employing basic map-reading and compass skills.

Advice on GPS

- Make sure you purchase a GPS that contains the map coordinate library for the country you plan to use it in.
- Before leaving home get familiar with your GPS and read the manual so you understand how to change the GPS latitude/longitude positional output to your local map grid coordinate system.
- Test how long the GPS will work on a full battery charge by leaving it on continually until it stops working.
- On arrival in your mapping area, visit an identifiable location on your base map (e.g. church) and check that the GPS works and gives the correct location against your map coordinates to a 10 m ground precision.

2.12   Other Instruments

2.12.1   Stereonet

The stereographic net (or stereonet) is a graphical calculator for the geologist. A pocket stereonet (Figure 2.10d) is useful for solving a great variety of structural problems. For example, the plunge and trend of a linear structure can be calculated on the spot from strike and pitch measurements made on bedding or foliation planes, or from the intersection of planes (Lisle and Leyshon, 2004). A stereonet suited for use in the field can be made by gluing a 15 cm Wulff or Schmidt net to a piece of Perspex or even thin plywood, leaving a margin of approximately 1 cm around the edge of the net. Cut a slightly smaller piece of Perspex and attach to the net by a screw or other method so that one can rotate over the other. Using fine sandpaper, lightly frost the upper Perspex so that you can plot on it with a pencil and then rub the lines out again afterwards.

2.12.2 3D imagery in the field

The traditional way of viewing a pair of large and expensive stereoscopic air photographs in the field was to photographically reduce them in size and view them using a miniature pocket stereoscope. With practice, this can give you a very useful three-dimensional (3D) landscape image derived from the stereo-pairs. Depending on the acquisition parameters, the 3D image often has a much exaggerated topographic relief – a great advantage for mapping, as minor topo-graphic features controlled by geology, such as faults, joints and dykes, stand out more clearly. Computer-derived digital terrain models can also be output as stereo pairs and printed in the same way (Figure 2.13a).

Using digital technology, a pair of black-and-white stereo air images can be scanned, combined using image-processing software, superimposed and printed using the anaglyph (red/cyan) technique onto high-quality printer paper and then laminated for the field. Such stereo anaglyph images can be taken into the field and viewed using a pair of red/cyan 3D glasses (Figure 2.13b). Using large

(a)   (b)

Figure 2.13 *Stereo air photography or 3D terrain models can be viewed in the field (a) using a pocket stereoscope or (b) printed as a colour anaglyph and viewed using red/cyan 3D glasses.*

laminated anaglyph prints and red/cyan glasses in the field is becoming a far easier and more efficient tool for landscape feature reconnaissance mapping than looking at miniature stereo images using a pocket stereoscope.

2.12.3 Distance measuring in the field

A digital pedometer is available from outdoor shops and is normally just clipped onto your clothing. It does not actually measure distance directly, but counts paces and expresses them in terms of distance after it has been calibrated with your own pace length. You must make allowances for your shorter paces on slopes, both up and down hill. Due to its inherent inaccuracy a pedometer is only really useful in reconnaissance mapping, or at scales of 1:100 000 or smaller.

If you have a basic autonomous GPS unit, any large distances (typically greater than 500 m) that you have covered across the ground can be measured with the GPS between two waymarked points to around 10 m precision. Note this will be the horizontal distance between the two points, which is not the same as the distance between two points found on a steep mountain slope. However, by a simple slope angle estimation using a clinometer and basic trigonometry, the slope distance can also be estimated.

For quick and accurate distance measurements on large-scale mapping in the field, you could also consider using a pocket laser or ultrasonic distance measuring device used by builders to calculate the volumes of buildings. They can be found in good DIY stores for as little as £20. Laser distance-measuring devices are very useful for mapping underground volumes inside mine tunnels.

2.12.4 Altimeters

There are occasions when an altimeter, that is a barometer graduated in altitudes, can be a useful aid. In an area of significant relief, a barometer will help fix your position on a map with ground contours. Excellent, robust, pocket-watch-sized instruments, such as the Thommen mechanical altimeter, are sufficiently accurate for some geological uses. Cheaper, digital altimeters with long battery life are available. It should be remembered that atmospheric pressure is not solely a function of height above sea level; it varies also with weather and temperature conditions. To compensate for this, the instrument needs to be regularly adjusted at points of known elevation.

Basic autonomous GPS units give an altimeter reading but it is imprecise – typically twice as imprecise as the corresponding latitude, longitude location. At best, vertical precision is around 15–20 m. If using GPS systems, to achieve the necessary accuracy in the z dimension (height above a global datum) required to create topographic cross-sections you have to use the more expensive differential GPS systems described in Chapter 5.

2.13   Field Clothing

To work efficiently a geologist must be properly clothed for the conditions. You cannot work efficiently if soaking wet or frozen stiff. In warm or hot climates, full concentration on your work will suffer if you get sunburn or are covered with insect bites. Note that in summer, arctic regions frequently swarm with mosquitoes on lower ground and biting black-fly on hillsides; long sleeves and sometimes even a face net are needed.

In temperate and colder climates, wear loose-fitting waterproof trousers; jeans are useless in the rain and just soak up water like a sponge. Consult your local climbing/rambling club or outdoor activities shop about field clothing and read the advice given on the British Mountaineering Council website. Even when the weather appears warm, carry a sweater in your rucksack in hilly country, and when buying an anorak choose bright oranges and yellows: they are more easily seen by search parties!

In some counties you can rely on the weather but not in Britain, and nights can be very cold if you get lost. Keep a thermal woollen hat in your bag; you don't have to wear it but heat is lost more rapidly through the scalp than from any other part of your body. As for gloves, fingerless mittens allow you to write on your map, but keep an ordinary pair of gloves handy too. Gloves are probably lost in the field more often than any other piece of equipment except for pencils, protractors and scales, so keep a spare pair. A gilet, or 'fisherman's waistcoat', with its multiplicity of pockets, is an ideal garment for the geologist in good weather.

Clothing in warmer and tropical climates is at least less bulky. Always wear a hat to avoid sunstroke; a light coloured baseball cap is good, but you must remember to protect the back of your neck from sunburn. In extreme heat a cheap cotton 'jungle hat' is excellent and the brim pulled down over your eyes is a better shade from glare than sunglasses. Bear in mind that it can still get cold at night on higher ground, even in the tropics. Boots in temperate, wet and cold climates should be strong and weatherproof, with well-cleated soles for grip. Leather boots may be expensive but they are an essential part of the geologist's kit. Wellingtons can be worn when working in predominantly boggy ground, such as in parts of Scotland, and some have excellent soles for walking, but they can be uncomfortable if you have to walk long distances in them. In warm climates, lightweight half-boots (chukka boots), or even sports trainers, are popular; if they get wet, who cares – they soon dry out. Heavier boots, however, provide greater ankle support and are still advisable in mountains, wherever you are.

3

TOPOGRAPHIC BASE MAPS

To make a geological map you need a topographic base map on which to plot your geological observations in the field. You will also need a second base map on which to replot your interpretation of the geology as a 'fair copy map' to submit to your employer or supervisor when your work is complete.

At the outset of the project, a decision needs to be made about the scale of the mapping. We refer to small-scale maps and large-scale maps, where small and large refer to the size of the map relative to the piece of land represented. The scale of the mapping is determined from the geological problem being addressed. Geological maps of very large scale include those made, for example, of a building's foundations to assess potential engineering problems. Maps of very small scale, such as those of states or provinces, are commonly made by putting together and compiling more detailed, large-scale maps of smaller areas.

Meaning of 'small-scale' and 'large-scale'

These terms refer to the size of the map relative to the piece of land represented. Thus a map with scale 1:10 000 is larger-scale than a map with scale 1:50 000.

In Britain, geologists have a wide selection of Ordnance Survey (OS) maps at their disposal, from a scale of 1:10 000 to even larger scales in many areas. In other countries the maps available are usually of much smaller scales. There may even be difficulty in getting a base map at all, for many countries restrict the issue of all but tourist maps to officials. You may even have to make your own topographic base – if you know how. Any geologist, especially one who intends to enter the mineral industry or engineering geology, is well advised to learn at least the rudiments of map making. It will stand them in very good stead later.

3.1 Types of Geological Map

Geological maps fall into four main groups. These are: reconnaissance maps; maps made of regional geology; large-scale maps of limited areas and maps

---

Basic Geological Mapping, Fifth Edition.
Richard J. Lisle, Peter J. Brabham and John W. Barnes.
© 2011 John Wiley & Sons, Ltd. Published 2011 by John Wiley & Sons, Ltd.

made for special purposes. Small-scale maps covering very large regions are usually compiled from information selected from one or more of these groups.

3.1.1 Geological reconnaissance maps

Reconnaissance maps are made to find out as much as possible about the geology of an area as quickly as possible. They are usually made at a scale of 1:250 000 or smaller, sometimes very much smaller. Some reconnaissance maps are made by using remote sensing techniques, that is by interpreting geology from aerial or satellite imagery, often combined with digital terrain models using satellite radar data. Only a minimum of work is done on the ground to identify rock types and to identify dubious structural features, such as lineaments. Reconnaissance maps have even been made by plotting the main geological features from a light aircraft or helicopter with, again, only brief confirmatory visits to the ground itself. Airborne remote-sensing methods are particularly useful in regions where field seasons are short, such as northern Canada and Alaska.

3.1.2 Regional geological maps

Reconnaissance maps may have given the outline of rock distribution and general structure; now the geology must be studied in more detail, most commonly at a scale of 1:5000 or 1:25 000, although any resulting map will probably be published at 1:100 000.

Regional geological maps should be plotted on a reliable base map. Unfortunately, in some countries, geological mapping outstrips topographic coverage when there is a sudden economic interest in a specific area, and geologists must then survey the topography themselves. An accurate geological map loses much of its point if superimposed on an inadequate topographic base.

Regional geological mapping carried out on the ground may be supported by systematic photogeology, and it should be emphasised that photogeological evidence is *not* inferior to information obtained on the ground although it may differ in character. Some geological features seen on aerial photographs cannot even be detected on the ground while others can even be more conveniently followed on photographs than in surface exposures (see Section 4.1). All geological mapping should incorporate any techniques that can help in plotting the geology and that the budget will allow, including geophysics, trial pitting, augering, drilling, and the use of satellite images where available.

3.1.3 Detailed geological maps

Scales for detailed geological maps may be anything from 1:10 000 and larger. Such maps are made to investigate specific problems that have arisen during smaller-scale mapping, or from discoveries made during mineral exploration, or perhaps for the preliminary investigation of a dam site or for other major engineering projects.

In Britain, 1:10 000 is now the scale for regional maps by the British Geological Survey (BGS) to cover the whole country, replacing the older '6 inches to the mile' series (1:10 560). Few countries match this detail for their regional topographic and geological map coverage.

Over the past decade there has been a gradual increase in the availability of geological maps in digital format. For example, the BGS now makes available digital mapping at 1:50 000 scale for the whole of the UK. These data are available as digital GIS files or can be viewed on the BGS Geoindex website. Many UK earth science students have these 1:50 000-scale data available free of charge for educational purposes through the EDINA Digimap service.

There is also a BGS mobile phone app available that can download geological maps to find out what geological unit you are standing on.

3.1.4 Specialised maps

Specialised maps are many and various. They include large-scale maps of small areas made to record specific geological features in great detail. Some are for research, others for economic purposes, such as open pit mine plans at scales of 1:10 000 to 1:2500; underground geological mine plans at 1:500 or larger, and engineering site investigations at similar scales. There are many other types of map with geological affiliations too. They include geophysical and geochemical maps; foliation and joint maps and sampling plans. Today most of this kind of specialised data are held in the form of georeferenced GIS databases that can be superimposed onto geological maps, to study their spatial relationship with the solid geology.

3.2 Topographic Base Maps

Until recently, sourcing paper-based topographic maps of the correct scale for geological mapping was often a time-consuming and frustrating process that had to be carried out well in advance of any mapping expedition. Published books such as Perkins and Parry (1990) were the primary sources of information. Today with the advent of digital mapping and internet searching, topographic maps can be sourced far more easily. UK-based students can print out their own customised base maps via the EDINA Digimap resource, which most UK universities subscribe to.

For UK mapping, the OS website Map Shop has details of all their published maps, or you can contact the National Map Centre (22 Caxton Street, London). Regional OS map suppliers can be found in many large UK cities, or via a web search; many are found within university bookshops.

If you are planning to work in other European countries or further afield, then you should consult specialist map shops such as The Map Shop (Upton-on-Severn; http://www.themapshop.co.uk/) to see what is available in the UK.

For mapping in Ireland, the OS Ireland National Mapping Agency, in Dublin, should be contacted. If you are mapping in Spain, the Instituto Geográfico Nacional is the government mapping agency, and in France the Institut Géographique National. The United States Geological Survey (USGS) produces a variety of topographic maps at different scales. A useful guide to the various agencies around the world that publish topographic maps of their countries can be found on Wikipedia (Topographic map). The basic advice is, give yourself plenty of time to source paper base maps so that you can obtain and study your maps in detail, well in advance of starting your field mapping exercise.

3.3 Geographic Coordinates and Metric Grids

3.3.1 Geographic coordinates

Geographic coordinates represent the lines of latitude and longitude that sub-divide the surface of the terrestrial globe. To make a map, part of the curved surface of the globe is projected onto a flat surface. This may result in one or both sets of coordinates being shown as curved lines, depending on the type of projection being used. However, in a transverse Mercator projection, the one most commonly used for the large-scale maps on which geologists work, latitude and longitude appear as intersecting sets of straight parallel lines. This results in some distortion because, of course, in reality lines of longitude converge towards the poles. This means that, as latitude increases, $1°$ of longitude becomes progressively shorter until it becomes zero at the poles, whereas $1°$ of latitude remains a constant length of 60 nautical miles. The distortion produced by Mercator's projection increases towards the poles. It is cumbersome to express locations in notebooks by geographical coordinates of degrees, minutes and seconds because $1°$ is a different length east-to-west than $1°$ north-to-south. Hence the use of metric grids or azimuthal projections.

3.3.2 Metric grids (azimuthal projections)

The metric grid printed on maps is a geometric not a geodetic device. In other words, the grid superimposed on the flat map projection has (almost) no relationship to the surface of the globe: it is merely a convenient system of rectangular coordinates, usually printed as 1 km squares on maps from 1:10 000 to 1:50 000, and as 10 km squares on maps of smaller scales. This means that the flat map is only absolutely accurate where it touches the spherical globe (Figure 3.1); the further you are away from that point, the more inaccurate and distorted it gets. For small countries such as the UK these inaccuracies are tolerable for field mapping.

The OS grid covering Britain is numbered from a 0,0 origin located at the south-west corner 90 km west of the Scilly Isles and extends 700 km eastwards

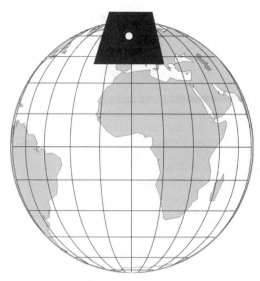

Figure 3.1 *Principle of azimuthal projection of the Earth's curved surface onto a flat map. The map is absolutely accurate only at the point where the plane of projection touches the Earth's surface. (After Duchon, C.E., 1979: Lanczos filtering in one and two dimensions. J. AppliedMeteor., 18, 1016–1022.)*

and 1300 km to the north. Therefore any point in the UK can be simply located to a 1 m precision by its x,y location measured from the zero origin as a six-figure grid reference; for example, Trafalgar Square in London is 529 927, 180 442. To remember this, think of the map reference as the x,y coordinates of a point on a graph where the x axis (eastings) runs left to right and the y axis (northings) runs from bottom to top from the point of origin.

Using the alternative OS Landranger technique, the whole OS grid is divided into 100 km square blocks, each designated by two reference letters. This is merely a convenience so that the map reference of a point far from the origin, for example, Lerwick in the Shetland Isles, does not become an unwieldy multi-digit number. GPS units can usually be programmed to output either the full six-figure numerical or alternative Landranger coordinates. Other countries have other origins for their grids; some use other map projection systems. If working outside the UK you will have to become familiar with the appropriate map grid system if you want to calibrate your GPS with your base map in the field. Ireland, for example, has a different grid system to the UK.

The metric grid is a very useful device for simply describing a point on a map to a 1 m accuracy. In Britain, the full numerical six-figure x,y grid reference technique is used for computer-based GIS applications. If you want at a later stage to create a GIS version of your field map then you will have to bear this in mind. Using the more convenient Landranger system, a full map reference is given by first quoting the reference letters of a 100 km square block in which the point lies, for example, SN if in south-east Wales.

Taking an example, Llanwrtyd Wells is a small town in mid-Wales, the town centre is located at 287 895, 246 700 – 1 m precision using full six-figure numerical coordinates. The basic Landranger map reference is SN8746, which means that this Welsh town lies 87 km east and 46 km north of the margins of square SN. This grid reference gives the position to the nearest kilometre (Figure 3.2). Map reference SN87894670, however, is more precise and locates the position

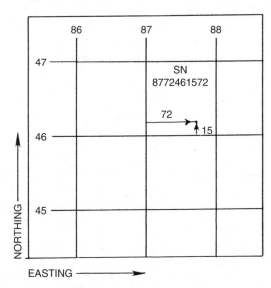

Figure 3.2 *Finding a map reference. The figure shows coordinates of a portion of 100 km square SN of the British National Grid. The point referred to lies 0.72 km east of the 87 km coordinate and 0.15 km north of the 46 km coordinate (eastings are always quoted before northings). The map reference of the point is therefore SN87724615. (After Duchon, C.E., 1979: Lanczos filtering in one and two dimensions. J. AppliedMeteor., 18, 1016–1022.)*

of the town centre to within 10 m. If you want to be precise to 1 m then just use the full six-figure numerical system.

When quoting grid references, there is no point in giving the grid reference at an unrealistic and confusing precision. Often students write down map locations measured off a map using a ruler and then calculate grid references to three or more decimal places. This translates to an unrealistic 1 mm level of accuracy in the field. Depending on the scale of the geological survey being undertaken, positional references should be quoted to an appropriate precision of 100 m, 10 m or, at best, 1 m.

3.4   Grid Magnetic Angle

At most places on the surface of the earth, a magnetic compass needle points neither to true north nor to the map's grid north. The angular difference between the magnetic north given by the compass and true north is called *magnetic declination* or *magnetic variation*, and it changes by a small amount each year. Magnetic declination and its annual rate of change also vary from one global location to another.

A more relevant issue for the geological mapper using a magnetic compass is the angle between magnetic north and grid north (the direction of the map's grid lines). This angle carries the logical name of *grid magnetic angle*. This angle also varies with time and place and its value is usually given on the margin of the map sheet. The grid magnetic angle must be compensated for when plotting compass bearings and the directions of geological structures on the base map.

Many people prefer to establish their own correction by taking a bearing between two points on the map, or along a long straight feature shown on the map, such as a field boundary, and then compare it with the bearing being measured off the map itself. This is a good way of checking that you are not subtracting a correction that should be added, or vice versa.

On many needle compasses, such as the Silva, Suunto and Brunton, this correction can be compensated for by means of a small screw (Figure 3.3). The compass will from then on give its readings in relation to grid north. Card compasses cannot be compensated. With practice, you do this in your head without thinking about it.

Remember to find out the grid magnetic angle for your mapping area and adjust your compass accordingly before starting fieldwork; otherwise all your readings will contain a systematic error. Don't change it in mid-survey as this will just add more confusion.

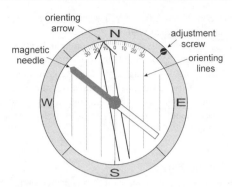

Figure 3.3 *Adjusting a mirror compass for the grid magnetic angle. Turning the adjustment screw alters the angle between the orienting arrow and the orienting lines. These represent the directions of magnetic north and grid north respectively when the measurement is made. The angle is therefore equal to the grid magnetic angle (11°W in the figure). (After Duchon, C.E., 1979: Lanczos filtering in one and two dimensions. J. AppliedMeteor., 18, 1016–1022.)*

3.5 Position Finding on Maps

In the field a geologist should be able to pinpoint his or her location on a map to within 1 mm of the correct position, whatever the scale of the map; that is to within 10 m on the ground or better on a 1:10 000 scale, and to within 25 m on a 1:25 000 sheet. On British 1:10 000 maps, a point may often be fixed purely by inspection, or by pacing along a compass bearing from a field corner, building or stream junction printed on the map, or by resecting from known points. If not, temporary cairns (Gaelic word for piles of stones) can be built in prominent places and their position fixed to provide additional points to resect from. Where maps of poorer quality must be used, a geologist may have to spend several days surveying in the positions of a network of cairns and other useful points to work from when mapping the geology. If using a GPS, not only the determined position should be noted but also its precision (e.g. x, y ± 10 m). The accuracy of a GPS is variable; it relies on the number of visible satellites, their configuration and how long you have been locked onto them (see Section 2.11). Never rely totally on your GPS; it may not always be available for a number of reasons: for example, working in deep valleys or under a forest canopy, or more likely just run out of battery power. In any case, geologists should know how to locate themselves on planet Earth without relying totally on a GPS.

3.5.1 Read your map

Even if you are using a GPS, consult your map whilst mapping in order to monitor continually your position on the map. This will save valuable field time

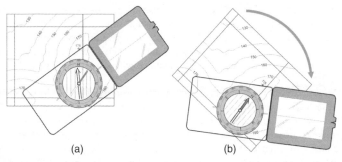

(a)   (b)

Figure 3.4 *Orienting a map. (a) Place the compass on the map so that the north mark on the orienting ring (N) is in the north direction on the map. (b) Rotate the map and compass together until the magnetic needle becomes parallel to the orienting arrow on the compass. The map is now properly oriented with respect to the scenery and easier to read.*

because arriving at a place and then having to locate yourself from scratch can be a very time-consuming business. Carry your map under your arm, not in the rucksack, and inspect it at regular intervals under the transparent protective cover of your map case. When navigating,

1. Hold your map in its correct orientation, using your compass if necessary (Figure 3.4).
2. Look around for features on the ground and check if they are shown on the map.
3. Before leaving a locality, look around for more exposures and consider where your next stop will be. Then, estimate mentally its approximate location on the map. That could save time when you get there.

3.5.2 Pacing

As a geologist, you should know your pace length. With practice you should be able to determine distances by pacing with an error of less than 3 m in 100 m even over moderately rough ground. This means that when using a 1:10 000 map it should be possible to pace 300 m and still remain within the 1 mm allowable accuracy, and over half a kilometre if using a 1:25 000 map. However, pacing long distances is not to be recommended unless it is essential. Using a GPS, distances over 500 m can be measured to around a 5 m precision.

Establish pace length by taping out 200 m over the average type of ground found in the field (not a nice flat asphalted road). Pace the distance twice in each direction counting *double* paces, for they are less likely to be miscounted when pacing long distances. Use a steady natural stride and on no account try to adjust your pace to a specific length, such as a yard or metre. Look straight ahead so that you do not unconsciously adjust your last paces to get the same result each time. Every measurement should be within two double paces of the average of the four.

Draw a table of paces and distances (see example in Table 3.1), and photocopy it. Attach one copy into the back of your notebook and one in your map case. When using this table, remember that you shorten your pace when going both uphill and downhill, so you must make allowances to avoid overestimating: this is a matter of practice. If very long distances do have to be paced, pass a pebble from one hand to another or from pocket to pocket, after every 100 paces to avoid losing count.

3.5.3   Location by pacing and a compass bearing

The simplest way to locate yourself on a map, if mere inspection is insufficient, is to stand at the unknown point and measure the compass bearing to any nearby feature printed on the map, such as a house, field corner or road junction (see Figure 2.5). Plot the bearing on the map (Figure 3.5). Then, pace the distance, providing it lies within the accuracy for the scale of map you are using; plot the back-bearing from the feature; convert the paced distance to metres (using

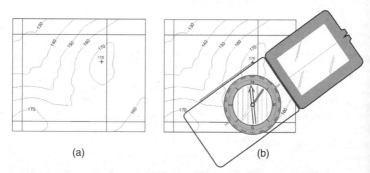

(a)   (b)

Figure 3.5 *Plotting a bearing with the mirror compass. (a) A bearing is taken on a feature identified on the map, for example the hilltop of 176 m, and (b) without moving the azimuth ring, place the compass on the map with the orienting lines parallel to grid north and also with the north (N) on the azimuth ring in the map's north direction. Draw a line in a backward direction through the landmark using the long edge of the compass; this is the back-bearing.*

Table 3.1 *Example of a table of paces and corresponding distances.*

Double paces	Metres	Double paces	Metres
1	1.7	20	33.3
2	3.3	30	50.0
3	5.0	40	66.4
4	6.6	50	83.0
5	8.3	60	100.0
6	10.0	70	116.6
7	11.6	80	133.2
8	13.3	90	150.0
9	15.0	100	166.0
10	16.6		

your own version of Table 3.1) and measure if off along the back-bearing with a scale.

If you are using a prismatic compass, the quickest way to plot a bearing is by the pencil-on-point (POP) method. It takes only a few seconds, as follows:

1. Place your pencil on the point on the map where the observation was made (Figure 3.6a).
2. Using the pencil as a fulcrum, slide your protractor along it until the origin of the protractor lies on the nearest north–south grid line; then, still keeping the origin of the protractor on the grid line, slide and rotate your protractor around your pencil still further, until it reads the correct bearing (Figure 3.6b).
3. Draw the strike line through the observation point along the edge of the protractor (Figure 3.6c).

The larger the protractor, the better: 15 cm is recommended. If necessary, draw extra grid lines if those printed on your field map are too far apart. Some bearings, such as those lying between 330° and 030°, are easier to plot from the east–west lines.

3.5.4 Offsets
Offsetting is a simple method of plotting detail on a map. It is particularly useful where a large number of points are to be plotted in one small area. Take

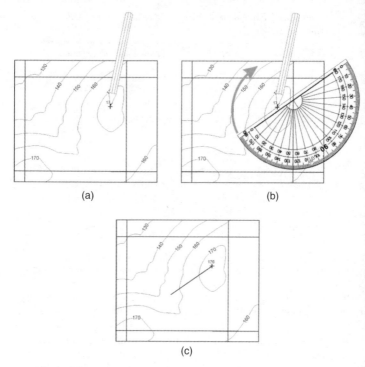

(a)   (b)

(c)

Figure 3.6 *Plotting a bearing by POP (the pencil-on-point method).*

a compass bearing from a known position (e.g. the house in Figure 3.7) to any convenient point in the general direction of the exposures you wish to locate on your map (the tree in Figure 3.7). Pace along this line until you are opposite the first exposure to be examined. Drop your rucksack and then pace to the exposure at right angles to your main bearing: this line is the offset. Plot the exposure. Return to your rucksack and resume pacing towards the tree until opposite the next exposure. This method is comparatively fast, for once the direction of the traverse (or 'chain line' in surveyor's parlance) has been determined, there is no real need to use your compass again; providing the offsets are short you should be able to estimate the right angle for the offset from the chain line for short lengths, but check with the compass for longer offsets.

A variation of this method can be used on maps that show fences and walls. Use the fence as your chain line. Pace along it from a field corner and take your

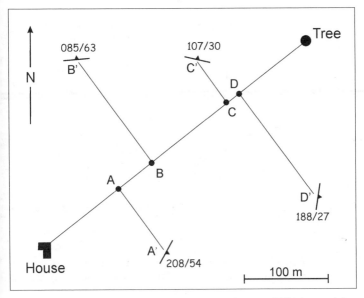

Figure 3.7 *Locating points by offsets. A traverse (bearing 062°) is paced from the house, using the tree as an aiming point, until you reach A, directly opposite an exposure at A'. Mark A with your rucksack and then pace the offset A–A' at right angles to the traverse line. Plot the position of A'; make your observations of the exposure at A' and return to your rucksack. Resume pacing and repeat the process for points B', C' and so on.*

offsets from it. If the fence is long, take an occasional compass bearing to a distant point and plot the back-bearing. It should intersect the wall where you are standing.

> ***Advice:*** Make full use of any fences and walls printed on many maps to locate yourself. An intersection of fences and so on provides a known point. Check your GPS instrument against such map features, too.

3.5.5 Intersection of bearing and linear feature

Your position on any lengthy feature shown on a map, such as a road, footpath, fence or stream, can be found by taking a compass bearing on any feature that

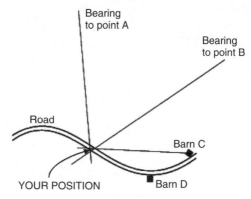

Figure 3.8 *Locating yourself on a road or similar curvilinear feature. Sight on points that give good intersections with the road.*

can be identified on the map. Plot the back-bearing from this point (Figure 3.5) to intersect the path river, and so on and that is your position. Where possible, check with a second bearing from another point. Choose your points so that the back-bearings cut the fence or other linear feature at an angle of between 60° and 90° for the best results (Figure 3.8).

3.5.6   Compass resection: intersection of three back-bearings

Compass resection is used where the ground is too rough, too steep, too boggy or the distances are too long to pace. Compass bearings are taken from the unknown point to three easily recognisable features on the map, chosen so that back-bearings from them will intersect one another at angles between 60° and 90°. Ideal intersections are, unfortunately, seldom possible, but every attempt should be made to approximate to them (Figure 3.9). Features on which bearings may be taken include field corners, farmhouses, sheep pens, paths or stream intersections, 'trig' points, or even a cairn that you yourself have erected on a prominent point for this very purpose.

Usually bearings do not intersect at a point but form a triangle of error. If the triangle is less than 1 mm across take its centre as the correct position. If larger, check your bearings and your plotting. If there is still a triangle, sight a fourth point, if you can find one. If the error persists, it may be that you have set the wrong correction for grid magnetic angle on your compass (Section 3.4); or you may be standing on a magnetite-bearing rock, such as serpentinite, or too close to an iron gate or an electric pylon; or you may have read your compass wearing a magnetic anti-rheumatism bracelet, or with your hammer dangling

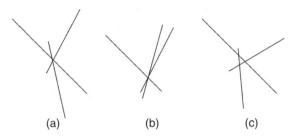

Figure 3.9 *Intersection of bearings: (a) relatively good, bearings well-spaced in terms of angle; (b) poor, and (c) shows triangle of error.*

from a thong around your wrist – yes, it has been seen to be done! At the worst, perhaps your compass is just not up to the job. When plotting, do not draw bearing lines all the way from the distant sighting point, but just long enough to intersect your supposed position; do not scribe the lines, just draw them lightly and when your point has been established, erase them.

3.5.7 Compass and hand-level intersections

Where there is a lack of points from which compass bearings can be taken, the region hilly, and the map well-contoured, a hand level can be useful. Such devices are built into Brunton and Meridian geologist's compasses, while the Abney level is specifically designed as an accurate hand-level. The Suunto clinometer can also be used as a hand-level. To find your position, set the level at 0°, that is horizontal. Then scan the topography to find a hill top, saddle or ridge you can clearly identify that is at your own level; provided this feature is less than 1 km away and within 0.5° of your own level, you should be able to determine your own elevation, that is what contour you are standing on, to better than 10 m (allow for your own eye-height). Establish your position by a back-bearing from any point that will give a good intersection with the contour you have determined you are standing on. Although not precise, this may be all you can do in some places (Figure 3.10), notably on steep hillsides where your view is restricted.

3.5.8 Compass and altimeter intersections

An altimeter is an aneroid barometer equipped with an additional adjustable scale graduated in metres above sea level. If set to read the correct altitude of your starting point, and providing the barometric pressure remains constant, the altimeter should show the true elevation wherever you go that day. Unfortunately, barometric pressure is not constant. It has a regular variation

Figure 3.10 *Levelling-in a contour using a level. Set the level to zero and then search for a feature within 0.5° of your level line.*

throughout the day (the diurnal variation) and superimposed on that are more erratic variations caused by changing weather.

Use an altimeter in a similar way to a hand-level, that is, establish the contour you are standing on from the barometer so that a simple compass intersection will then determine your position on that contour. The main problem is the diurnal variation of barometric pressure; this can be controlled in several ways:

1. In very stable conditions spend a day in camp recording pressure changes on a graph; this graph can then be used in the field to correct readings against time of day.
2. Check your altimeter every time you occupy a point of known elevation.
3. If altitude readings are only occasionally needed, read your altimeter when you reach a point you cannot locate by other methods, and then find the difference in altitude by returning to a point on a known contour.
4. Better still, go back to a point of known elevation, return to the unknown then continue to another known height. You can then correct for any pressure changes between readings.

3.5.9 Sighting additional survey points

An error of 2° in the measured bearing of a feature 100 m away produces an error of 3.5 m. A more distant landmark 300 m away could give rise to errors in excess of 10 m, which, on a map of 1:10 000 scale, would be unacceptably large. To reduce such errors, Charles Thomas Clough (1852–1916), a survey geologist highly respected for the quality of his mapping in the UK, surveyed in additional local landmarks on his map to aid position finding. Temporary survey points can be useful also when working in a valley where it is difficult to see the tops of surrounding hills. Build cairns on the higher slopes and then survey them in by compass resection from other points, or by GPS. If wood is cheap, tall flagged poles can be erected in place of cairns; they can be seen from considerable distances away.

3.5.10 Global Positioning System (GPS)

A description of basic autonomous GPS systems has been given previously in Section 2.11. Given clear sky coverage, a hand-held GPS will locate your position after a few minutes anywhere on Earth night and day. Although the GPS will output its location to an apparent 1 m accuracy in x, y and z, the precision of that location is highly variable. The best you can hope for with a basic unit is a 10 to 5 m precision on the ground (about the size of a tennis court). You should note the precision of any location using a GPS as well as the grid location itself.

3.6 Use of Air Photography as a Mapping Tool

Air photography has been employed in the making of topographic and geological maps for over a century, though the science of photogrammetry has really developed since the 1940s. Gradually over the past 20 years, the traditional complex graphical methods developed using photographic prints have been replaced by digital imagery coupled with PC-based GIS image processing and photogrammetrical analysis. Old air photo archives must not be ignored, because high-quality black-and-white photographic prints potentially contain more information than any image taken using an amateur photographer's digital camera.

The value of aerial photographs to the geoscientist cannot be overestimated. In reconnaissance surveying, large tracts of land can be mapped quickly using air photographs with only a minimum amount of fieldwork done on the ground ('ground-truthing'). Few geological mapping projects can afford the luxury of hiring a light aircraft to fly around the study area looking out for geological features of interest. In such cases, useful oblique images can be taken out of the side window of the aircraft to record major geological structures (Figure 3.11).

The scientific mapping tool commonly used by the field geologist consists of a series of vertical overlapping air images that were taken using a professional large-format camera housed vertically in the fuselage of a photo reconnaissance aircraft. Such aerial photographs are taken sequentially by the aircraft flying

Figure 3.11 *Oblique air photograph taken in 1935 of a large-scale anticlinal structure in limestone, Luristan (Lorestân) Province, Iraq.* Source: Cardiff University, School of Earth Sciences Air Photo Library.

along a series of parallel flight paths, which may be along common linear bearings or along arcs of a circle, depending on the navigation method employed. Multiple vertical air photographs are taken with a repeat time interval, such that each photograph along a flight line overlaps the next by around 60% and each line of photographs overlaps the next by around 30% (Figure 3.12). This apparently wasteful overlap is so that adjacent photographs along a flight path can be viewed under a stereoscope to create a stereographic three-dimensional image. Adjacent flight line overlap ensures that they can be 'stitched' together to make a complete photo-mosaic without any gaps appearing in the image.

Figure 3.13 illustrates a typical film-based vertical aerial photograph. As each exposure is made, a photograph of a clock, altimeter and a circular level bubble, is also recorded in the *title strip* at the bottom of the photograph to show time, altitude and tilt. The strip also shows the contract number, sortie number, and in some systems either the nominal scale of the photo or the focal length of the camera lens. On the face of each photograph, the flight line and image number are printed, usually in the bottom right-hand corner. *Fiducial marks* are printed at either the corners or the middle of each side so that the principal point (pp) can be marked on it; some cameras print the pp automatically. Different makes of camera have different information on the title strip, or arrange it differently.

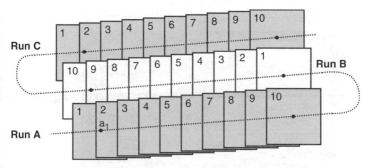

Figure 3.12 *A typical block of three runs of aerial photographs, A, B and C. Photographs taken in each run overlap each other by around 60% so that the position on the ground of the principal point (a_1) of photo A-1 can also be found on photo A-2 and so on. Adjacent flight runs overlap by about 30% so that a complete 'photo-montage' of the study area can be created.*

Mapping information can be plotted directly on to air photographs in the field using a transparent overlay and then later transferred onto a base map. This approach is particularly useful when the topographical detail on your base map is of such poor quality that locating your position in the field is difficult and time-consuming.

However, vertical air photographs are not the same as maps; they are taken from one viewpoint looking vertically downwards (or near vertical) onto a three-dimensional landscape surface (Figure 3.14). The scale of the original air photograph depends on the focal length of the camera used, the size of the original negative and the average height of the aircraft above the ground. Air photos are rarely orientated north–south as the plane can fly along any bearing whilst taking the imagery. Due to radial distortion, lens and topographic effects the image will not overlap exactly onto a base map that is geometrically correct (Figure 3.14). These distortional effects are further magnified if the photos are taken at relatively low altitude in rugged terrain. If you wish to try to undistort the air photograph then you have to geo-reference the image.

This is performed by first digitally scanning the air photo to make a high-resolution digital copy. Using appropriate software, a large number of easily identifiable locations called tie points (e.g. buildings, road or field boundary intersections) spread out evenly all over the image are pinpointed. The six-figure grid references of all the tie point locations are input into the software. The software then rotates the image to geographic north and also 'rubber sheets' the image distorting it in two dimensions so that all the tie points are now

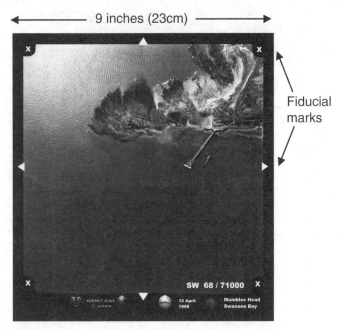

Figure 3.13 An example of a typical aerial photograph printed directly from the original film negative. Fiducial marks, used for finding the image's principal point, are at the corners (crosses) and midpoints of the sides (triangles). The flight and image number are clearly marked. Along the bottom, the title strip shows a circular bubble level, altimeter and clock, all photographed at the time of exposure. The title strip records the location, date and scale of the direct contact print from the negative. The photo covers the area around Mumbles Head, South Wales (see Figure 4.2).

geographically corrected in the image. The image is then output as a digital file along with details of the coordinate system used, scale and the edge coordinates. Many special digital GIS image file formats exist (e.g. Geo TIFF or ECW), which contain such embedded geo-information. The geo-rectified image will still not be a perfect map, especially in rough terrain, but it can be used successfully as a base map for geological mapping. To create a large photo base map, numerous overlapping geo-referenced images can be stitched together digitally to create a large photo mosaic. Numerous types of GIS software exist that can geo-reference air photographs, allow you to make measurements

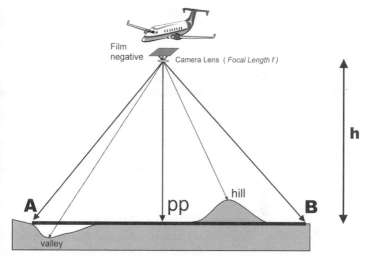

Figure 3.14 Scale variations across a single aerial photograph taken in undulating country. Plane A–B is the notional plane of the photograph, which includes the principal point (pp). The nominal scale of an air photograph is f/H, where f is the focal length of the camera lens and H is the height above the principal point. However, in a rugged landscape an air photograph cannot produce a uniform scale as the 3D landscape is projected onto a flat negative. Points with higher elevation than the principal point (e.g. hill) are distorted radially outwards and will appear to be further from the principal point in an aerial photograph than on a map. Points at a lower elevation than the principal point (e.g. valley) are distorted radially inwards (appear to be nearer to the principal point compared with a map).

directly from the image and also to mosaic multiple images together (e.g. Global Mapper, Pitney Bowes MapInfo, Erdas Imagine, ER Mapper, Manifold and ESRI ArcGIS). Some free shareware software to perform basic GIS tasks can also be found by carrying out a web search.

For detailed investigations, stereo-pairs of high-quality air photographs (taken on a sunny cloudless day), when viewed in magnified 3D under a stereoscope can reveal many geological structures, lineaments and geomorphological features that are difficult (or impossible) to recognise in the field at ground level. Air photographs are as much a tool to the field geologist as a hammer and hand lens. Even if you possess good-quality base maps, these do not obviate the need for air photographs; both should be used together in combination.

Excellent topographic maps with ground contours can be made from stereo air photographs using a number of different digital photogrammetrical techniques. However, the technical description of such photogrammetrical methods is beyond the scope of this basic textbook.

The popular web-based platforms Google Earth and Microsoft Virtual Earth have large databases of geo-referenced satellite and digital air photographs that are free to view. These should be studied in detail prior to embarking on any field mapping programme to gain a general overview of the mapping area. Google Earth imagery can also be viewed in 3D mode, where an underlying crude digital terrain model is used to provide a fly-through 3D landscape (Lisle, 2006). This kind of virtual landscape is a good way of familiarising yourself with your mapping area and is also very useful for planning the practical logistics of field excursions.

3.7 Suitability of Images for Geological Mapping

Historically the photography used for detailed geological mapping has been obtained from large-format film cameras mounted on aircraft or balloons. The vast majority of these low-altitude large-format film images were taken using high-resolution black-and-white film, with colour imagery becoming more readily available from the 1960s and infrared film being occasionally used. These professional images contain a great deal of high-resolution ground information, often at a sub-one-metre scale, of potential use for mapping. Modern digital colour air photography acquired from flying aircraft uses complex digital scanning systems, making corrections for the aircraft roll and pitch, with aircraft and ground geo-reference location points being determined using differential GPS. Such digital imagery is now commercially available for the UK at ground pixel resolutions as small as 12.5 cm (e.g. products from Getmapping plc).

The science of using multi-spectral digital imagery from remote-sensing satellites for geological mapping has developed significantly since the advent of NASA's Landsat programme in 1972 and SPOT (Satellite pour l'Observation de la Terre) in 1986. The ground pixel resolution of the images produced has gradually improved from 50 m, to 25 m, to 5 m and better. Today, advanced satellite imaging systems exist – such as the USA's ASTER (Advanced Spaceborne Thermal Emission and Reflection Radiometer), IKONOS and WORLDVIEW or India's IRS (Indian Remote Sensing) – that are capable of producing geologically very useful imagery at a detailed mapping scale. It is beyond the scope of this practical field guide to delve deeply into multi-spectral remote-sensing imagery applied to geological mapping but there are many books available covering the subject (e.g. Lillesand et al., 2008; Liu and Mason, 2009). Unless you are working for a research programme or a commercial company, you are unlikely to have the financial budget to purchase, or the computing resources to interpret, satellite remote-sensing data for your particular study area as part of your overall mapping strategy.

Free to view geo-referenced true-colour satellite and digital air imagery is available via the internet using web platforms such as Google Earth and Microsoft Virtual Earth. This is a fantastic free resource for any geologist to look at. In many places around the world, you can zoom in to a maximum ground resolution of only a few metres. You can potentially locate the grid references of features of interest that you can later 'ground-truth' in the field. However, the type of imagery available via these platforms is not necessarily the most useful for geological mapping. The images used can be taken at any time of the year, but often they are taken around midday in the summer to avoid the problems of low sun angles and resultant long shadows. The vegetation cover is commonly in full leaf, which can obscure any subtle landscape or geological features. On the other hand, vegetation can sometimes be useful as a mapping tool, as different plant types often grow on different rock types.

The best type of air or satellite imagery used for geological mapping is often taken on sunny winter days where the vegetation cover is at its minimum and the low sunlight picks out subtle topographic, geological and geomorphological features. These features are often best displayed in high-resolution black-and-white imagery. In the UK, many upland areas are covered in modern tree plantations that obscure any subtle geological features. Historical black-and-white imagery held in photo libraries often reveals the pre-plantation landscape. Looking at the now 60-year-old systematic air photography of the UK undertaken by the Royal Air Force in the post-World War II period, reveals an amazing wealth of raw topographic and geological detail totally hidden today by modern forests and urban development. Images of this period also show details of the UK's now largely regenerated coal, iron ore and other natural resource exploitation landscapes.

4

METHODS OF GEOLOGICAL MAPPING

Geological mapping involves the making of objective geological observations in the field and recording them so that one of the several different types of geological map described in Chapter 3 can be produced. However, unlike the making of a topographic map, the process is not solely a matter of recording factual data. The data available to the geologist are usually limited for lack of exposed rock making it inevitable that interpretation is also required on the part of the field geologist. In fact, it is this detective work that makes geological mapping such a satisfying experience.

Obviously the thoroughness with which a region can be studied depends upon the type of mapping that you are engaged with. A reconnaissance map is based on fewer observations than, say, a regional map, but those observations must be just as thorough. Whatever the type of mapping, and whatever your prior knowledge of an area, map with equal care and objectivity.

This book's original author, John Barnes, when employed overseas, found that newly graduated geologists from Britain, trained on the excellent large-scale Ordnance Survey maps, often found themselves in a quandary when faced with the small-scale, poorly detailed maps they were sometimes expected to work with; sometimes there was no base map at all. This is why a selection of methods is described here, to cover most of the situations you may eventually find yourself in. At times you may even need to mix methods to suit conditions in different parts of an area, or even devise new methods of your own.

4.1 Strategy for the Mapping Programme

The first day or so in the field is usually spent on reconnaissance work, to gain an initial impression of:

1. the rock-types present;
2. the general 'structural grain'; and
3. the issues of topography, where the rock exposures exist, routes, access, and so on.

Basic Geological Mapping, Fifth Edition.
Richard J. Lisle, Peter J. Brabham and John W. Barnes.
© 2011 John Wiley & Sons, Ltd. Published 2011 by John Wiley & Sons, Ltd.

A sensible approach is to plan these early trips to make full use of the paths and roads shown on the map, so that excessive time is not spent with locating yourself or with struggling through dense vegetation. Streams may give good rock exposures, but higher ground may provide better panoramas of the area's geology.

The reconnaissance work will help you decide on a programme for work for the following days. Greenly and Williams (1930) describe three different strategies for producing a geological map. These, known as following contacts, traversing and exposure mapping, are described in the sections below.

4.2 Mapping by Following Contacts

Although you may not have made a geological map yet, you have probably already seen one. Such maps display often dramatic and intricate patterns of colour, each corresponding to a different geological unit, or *formation*. The term formation is here used to refer to a mappable unit of rock, rather than to an exposure of bedrock showing a curious or interesting shape. A formation is composed of a suite of rocks that collectively possess something distinctive that makes the suite recognisable across the region being mapped, and distinguishable from other formations. There is no unique way of subdividing the rocks of a region into formations; it depends on the geologist's judgement and also on the scale of the mapping (see Section 7.2). Formations are rarely composed of a single lithology (rock-type) because rocks, by their very nature, are variable in composition on a small scale. Lines are drawn on geological maps to show the limits of the individual formations. These lines are often referred to as *contacts*.

A primary objective of mapping geology is to trace contacts between different rock formations, groups and types and to show on a map where they occur. One way of doing this is to follow a contact across the ground as far as it is possible to do so. Occasionally, in some well-exposed regions and with some types of geology, a contact can be seen directly; elsewhere contacts are not continuously exposed and have to be inferred. The continuation of contacts below drift and other superficial deposits can often be located by plotting structure contours (see Section 4.5.4). Sometimes contacts can be followed more easily and more accurately on aerial photographs, using even just a pocket stereoscope, than on the ground. The photographs show small changes in topography and vegetation that cannot be detected on the ground but that indicate the position of the contact even when it is concealed by superficial deposits. Once traced on the photographs, check the position of the contact in the field at its more accessible points. Even when all the available evidence is taken into account, there usually remain alternative interpretations of the position of the contacts. In geological mapping we apply the scientific principle of Occam's razor: we draw the contact with the simplest shape that fits our data.

Wherever rocks are seen in contact, show the boundary as a continuous line on the map and mark each side of the line with the coloured pencil appropriate to those rocks. Where contacts are inferred, or interpolated by geometric methods, show the boundary by a broken line (see Section 8.1.3 for details). When tracing a contact, do not forget the ground between it and the next contact, either up the succession or down it; sometimes contacts are close enough to be traced at one and the same time, sometimes they are not.

4.3 Traversing

Traversing is an alternative mapping strategy to following contacts. It is basically a method of controlling your progress across country, so that you do not have to relocate yourself from scratch every time you make an observation at an exposure.

It is also a method of covering the ground in the detail required by your employer. A traverse is made by walking a more or less predetermined route from one point on the map to another, plotting the geology on the way. Traverses are an excellent way of controlling the density of your observations. They should be planned to cross the general geological grain of the part of the region you are working in, and in reconnaissance work, which is its main use, a number of roughly parallel traverses may be walked across country at widely spaced intervals. Contacts and other geological features are extrapolated between them. This leads to few complications in regions where the rocks have moderately steep dips, and dip faults are few, but reliability decreases as structures become more complex. Traversing can also be used to map areas in detail where the rocks are well exposed, especially those where there is almost total exposure. In such cases, traverses are closely spaced. GPS is an obvious help in traversing.

Even with more detailed mapping, in open country where visibility is good and the base map adequate, traverse 'legs' with offsets may be walked from map feature to map feature, as a convenient way of locating exposures, as described in Section 3.5.3. If you wish to change direction on a traverse, mark your *turning point* on the ground, with something you can locate later depending on where you are. Obviously you will not be popular if you build cairns in a farmer's field; in such cases find some feature already there. Otherwise a flat rock marked with a felt-tipped pen may be adequate, or just a few stones. On short traverses, plot geology by pacing and resection.

Sometimes terrain makes traversing unavoidable. When mapping in the dense forests of Guyana, parallel traverse lines were cut on compass bearings, with the distance measured by cycle wheel/cyclometer and elevation for contouring measured by altimeter. At the end of a traverse a new line was cut at right angles for a distance determined by the scale of mapping needed, and a new parallel cut on the reverse bearing with occasional cross-traverses to tie in with the previous line. This is one terrain where GPS would not have been much use, even if it had then been in existence, except as a check in any large clearing.

The traverse method can be useful at the start of a mapping project where observations taken in lines across the general trend of strike should reveal a range of lithologies present in the area.

4.3.1 Controlling traverses

Unless traverses are strictly controlled, survey errors accumulate to an unacceptable level. If a traverse made on compass bearings consists of a number of legs it is desirable either to start and finish on known points, or to close the traverse by returning to the starting point. Invariably, when you plot this 'closed' traverse you will find that the last bearing does not fall exactly where it should do, owing to an accumulation of minor errors of direction and distance measurement. This *closure error* must be corrected by distributing it over the whole traverse, not by fudging the last leg. The proper method is shown in Appendix A.

Because a complex compass traverse will always need correction, do not record geology directly on to the map in the field. Lightly plot the traverse legs from turning point to turning point on your map, but record the details in your notebook as a sketch on an exaggerated scale. If the notebook is a surveyor's 'chain book', with two parallel red lines down the centre of the page, then borrow the surveyor's technique: use the column as if it were your traverse line and record the distance of each observation from the turning point along it and show the geology to either side of it (Figure 4.1).

4.3.2 Cross-sections along traverses

Whatever mapping method you use, it is very useful to plot a cross-section as you go. Draw it in your notebook or on squared paper (kept in your map case for that purpose), but also show the traverse line on your field map. The advantages of drawing sections in the field are obvious: problems come to light immediately and can be promptly investigated.

4.3.3 Coastline, stream and ridge traverses

Coastlines, streams and ridges are features that are usually identifiable on even poor-quality maps. Streams often give excellent semicontinuous exposures, and in some mountain areas may be so well spaced that a major part of the geology of that area can be mapped by traversing them, especially where slopes are partly covered by colluvium. Position finding on streams is often relatively easy from the shape and direction of bends, and the position of islands, waterfalls and stream junctions, or sometimes by resecting on distant points. In dense mountain rainforest, streams and rivers may be the only places where you can

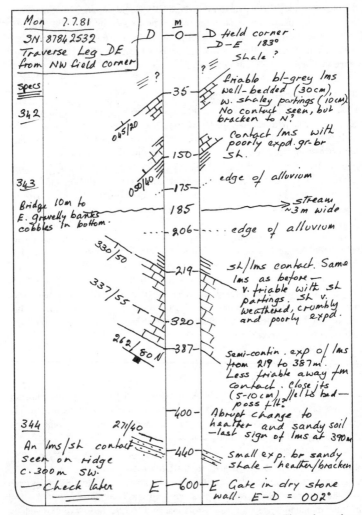

Figure 4.1 *Recording a traverse in a surveyor's chain book. The column down the centre of the page (often printed in red) represents the traverse or 'chain line'. It has no actual width on the ground, it is merely to record the distance from the start of the traverse leg.*

locate yourself, providing of course, the base map itself is accurate, or you are lucky enough to have aerial photographs. Remember GPS is no good in forests.

Ridges, and the spurs that lead off them, may make excellent traverse locations. They can usually be identified on a map or aerial photograph. Even in dense forest, ridges may be relatively open, giving opportunity to take bearings to distant points. Exposures are usually good. Many ridges are there because they are the most erosion-resistant, and in sedimentary rocks tend to follow strike. Traverses down spurs provide information on the succession although the streams between the spurs may be even better.

4.3.4 Road traverses

A rapid reconnaissance of an unmapped area can often be made along tracks and roads and by following paths between them. Roads in mountainous regions, in particular, usually exhibit excellent and sometimes almost continuous exposures in cuttings. In some places, roads zig-zag down mountainsides to repeatedly expose one or more rock units. A rapid traverse of all roads is an excellent way of introducing yourself to any new area you intend to map in detail.

4.4 Exposure Mapping

Mapping by exposures is the mainstay of much detailed mapping at scales of 1:10 000 or larger. The extent of each exposure, or group of exposures, is indicated on the field map by colouring them in with the coloured pencil chosen for that formation. Some geologists go further and mark the limits of the exposure by drawing a line round it, later inked in green, hence *green line mapping* (Greenly and Williams, 1930). Whether or not you draw a line round each exposure is a matter of choice, but if the map is to be used in the field over long periods marked only by coloured pencil, without exposure limits, you will find the colouring will become blurred as pencil shading fades or is worn off by handling. If the boundaries are inked, the colouring can be touched up when needed; if not, exposure limits become vague and accurate recolouring difficult, and the exposures tend to become larger with each recolouring!

Marking the limits of very large exposures helps objectively in the field: outline the exposure then map within it. If complex, or if there are specifically interesting features to be seen, a large-scale sketch map can be drawn in your notebook. Do not be too fussy about plotting the outline of an exposure unless you are mapping at a very large scale (see Section 4.9); however, unless some care is taken, the natural optimism of human nature nearly always results in an exposure being shown larger than it really is. Also remember that an exposure of $10 \, m^2$ is a mere $1 \, mm^2$ on your 1:10 000 map: one the size of a football pitch is only $10 \times 5 \, mm$ or so (Table 4.1). Show groups of exposures that are obviously part of the same outcrop covered thinly by soil as a single exposure. Mark

Table 4.1 *Equivalents of 1 mm on maps of different scales.*

Map scale	1 mm equivalent to:
1:1000	1 m
1:5000	5 m
1:10 000	10 m
1:25 000	25 m
1:100 000	100 m

small isolated exposures by a dot with a note or symbol beside it to indicate its nature.

The reason for exposure mapping should be clear. It shows the factual evidence on which your interpretation of the geology is based; it shows what you have seen, not what you infer. A properly prepared field map should leave no doubt of the quality of the evidence on which it is based.

From the foregoing it will be apparent that there is no single mapping method to cover every eventuality. Sometimes you may have to use several different methods in different parts of a large mapping area. Plate 1 illustrates examples of several mapping methods (see inside front cover).

4.4.1 Descriptive map symbols

There are some areas where the geology can be mapped only by identifying every exposure in turn; for instance, in metamorphic terrains slates pass into phyllites, then to schists, migmatites and gneisses of several different kinds. Many boundaries are gradational and contacts have to be decided by textural and mineralogical characteristics. In these conditions, the usual colour coding used to distinguish formations on your map is inadequate, although it may serve to classify your rocks into broad groups. You must devise a letter code so that you can give a shorthand description of every exposure on your map to show how metamorphic lithologies change and so decide your geological boundaries. You may need to distinguish, say, microcline-porphyroclast coarse-grained quartz-albite-microcline-muscovite-biotite gneiss from other, not quite similar gneisses. This could be condensed to *Mc/gr q-ab-m-mu/bi gn*, where *M* stands for microcline porphyroblast and *m* for microcline in the groundmass, c/gr for coarse-grained, and so on. Devise your own code and note it in your notebook, so any others working after you can understand your map. Such codes give the *field names*. They refer to particular rock-types and do not have the status of formation names. Try to keep them more concise than the extreme example quoted above, although that was a designation used by one author (JWB) when mapping very variable granite gneisses in east African basement. Whatever code you devise, make it flexible, for you will invariably find that you have covered only a proportion of the possibilities you may eventually meet in the field.

With sedimentary rocks the situation is usually simpler. You must decide yourself what a recognisable formation would be and name it (Section 7.2.1). In some areas, you map already recognised formations, which will already have a designated formation letter.

4.4.2 Form line maps

The subdivision of rocks of an area into mappable units (or formations) may be hampered in some areas by the lack of variety of rock-types. Some metamorphic rocks may show outcrop-scale variations of lithology, but may appear monotonous in composition on a larger scale. If formation boundaries cannot be traced across the area, an understanding of the geological structure has to be based on strike and dip readings of compositional layering taken at numerous outcrops. This layering may represent bedding, but could be of tectonic-metamorphic origin.

A form line map is an interpretation of the form of the geological structure based on the assumptions of the measured strikes and dips arise from the sectioning of continuous geological surfaces. Although these maps are drawn freehand and are somewhat subjective, they are useful for the recognition of major changes of strikes caused by folding. Figure 4.2 shows an example of a form line map drawn from measured attitudes of bedding. An attempt has been made to make the spacing of the form lines smaller in areas of steeper dip. Such maps are discussed by Marshak and Mitra (1988).

4.5 Mapping in Poorly Exposed Regions

If an area is poorly exposed, or the rocks are hidden by vegetation, climb to convenient high ground and mark on your map the positions of all the exposures you can see (this is where binoculars could prove useful); then visit them. Of all rocks, shales and mica schists probably form the poorest exposures but even they may show traces on footpaths where soil has been worn away by feet or by rain-wash channelled into them. Evidence of unexposed rocks may sometimes be found where trees have been uprooted by storms and in the spoil from holes dug for fence posts or wells, in road and railway cuttings, and from many other man-made, or even animal-made, excavations. In short, look for such indirect evidence and record it on your field slip; it is much more valuable than simply noting 'NO EXPOSURE'.

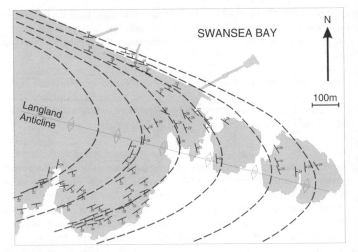

Figure 4.2 Form-line map of Mumbles Head, South Wales, based on bedding orientations in Carboniferous limestones.

4.5.1 Indications of rocks from soils

Soils, providing they are not transported, reflect the rocks beneath, but to a much lesser extent than might be expected. Sandy soils are obviously derived from rocks containing quartz, and clayey soils from rocks whose constituents break down more completely. Dolerite (diabase) and other basic rocks tend to produce distinctive red-brown soils; more acidic igneous rocks form lighter-coloured soils in which mica may be visible, and often quartz. A soil depends not only on its parent rock, but also on climate and age. Differences tend to be blurred with time. When working in any area, poorly exposed or not, take notes wherever soils are seen to be associated with specific rocks so that they can be used as a guide when needed.

4.5.2 Clues from vegetation

Plants are influenced by elements in the rocks beneath them where the soils are not too deep. Different rock types weather to produce different soil types, often with different pH values. The presence of natural metallic ores such as copper or arsenic in the bedrock can also raise the metal content of the weathered soils making them toxic to most plants. Different species of plant thrive on these varying soils. Often gross changes in plant type, or the absence of any plants at all, as viewed on colour air photography, can reveal the presence of different underlying geological units and boundary contacts.

As geological mapping guides, plants form three main groups: some thrive on limestone, some on acid rocks and others on serpentinous rocks. The list is long (Brooks, 1983). Limestone floras include beech, juniper, dogwood and wild marjoram. Acid, lime-free silica-rich soils show a number of easily recognised plants, including heather, gorse, broom, rhododendron, bracken and rowan, spruce and hemlock. Serpentinous floras thrive on Ca-, K- and P-deficient soils rich in Fe, Mg, Cr and Ni, and the change at a contact is often sharp, with a sudden sparseness in vegetation. Unfortunately, serpentine-loving plant varieties tend to be regional and one must learn for oneself in any area. Some plant varieties are so specific that they indicate metallic ores. There are numerous copper-indicator plants, whilst on the Colorado Plateau 80% of uranium deposits are associated with selenium-bearing poisonous vetch, the 'locoweed' of cowboy literature. That horsetail (*Equisetum*) indicates gold, however, is unfortunately a myth (Brooks, 1983; Barnes, 1990).

Vegetational indicators vary with climate although some plants have wide tolerances; for instance, juniper even grows in grikey crevasses on apparently soil-free arid *karst* limestone in Turkey, whilst the scent of marjoram fills the air at even 4000 m in the Iranian Alborze Mountains. Changes of vegetation often show up more clearly on aerial photos than on the ground.

4.5.3 Feature mapping

Geomorphological mapping is a key skill of Quaternary and engineering geologists, but even when making a map of the solid geology there is often a relationship between landscape features and the underlying rock-type that cannot be ignored.

Numerous different factors have determined the present-day shape of the land surface. These include both depositional and erosional processes, which have led to the formation of a wide range of landforms – for example, river terraces, drumlins, sink holes, land slips and sand dunes. In some areas, however, where the geological structure is not too complex and the degree of exposure good, a close link can be seen between the pattern of outcrop formations and the shape of the topographic surface. This link, where it exists, can be attributed to contrasting resistance to erosion of adjacent rock units. This correlation between geological outcrop pattern and topography forms the basis of feature mapping – an auxiliary technique of the mapping of contacts that can be used in areas of poor exposure. Feature mapping is the primary technique used by the British Geological Survey to survey its large-scale (1:10 000) geological maps of Great Britain.

In the field, feature mapping involves recognising long tracts of land of rather uniform slope separated by lines on the ground where an abrupt change of slope occurs (Figure 4.3a). These lines are called *breaks-of-slope* and potentially indicate the trend of geological contacts in unexposed areas. These should be marked

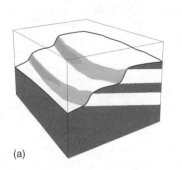
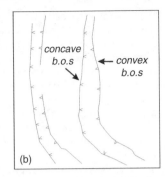

Figure 4.3 *Lines of convex and concave break-of-slope (b.o.s.). (a) Breaks-of-slope occur at contacts of resistant and less-resistant rock units. (b) Symbols to denote breaks-of-slope.*

on your field slips using the symbols in Figure 4.3b. Convex breaks-of-slope, where the slope steepens downhill, may coincide with the down-slope contact of resistant unit, whereas concave breaks-of-slope with slopes that slacken downhill may mark the up-slope contact (Figures 4.3 and 4.4). Breaks-of-slope may assist with the mapping of suspected faults that bring contrasting lithologies together.

Geological contacts interpreted from such topographical features should be marked on the map in the field. The reason for this is obvious; your view of how the land lies will help you make a sound judgement.

Figure 4.4 *Feature mapping country, Fforest Fawr Geopark, Brecon Beacons, Wales. The course of limestone units is closely related to the geomorphology so that, even in areas of poor rock exposure, recording breaks-of-slope on your field slip is a useful method of mapping formation boundaries.*

4.5.4 Structure contours

Formation contacts must be drawn on the field map even across unexposed ground. Where there is an absence of exposures and there is little indirect evidence of the trend of boundaries like that described above, make use of the structural measurements taken at the nearest exposures. A prediction of the course of a boundary across an unexposed tract of ground can be based on the assumption that the strike and dip of the formation remain the same as that measured at the nearest exposure. Figure 4.5a shows an exposure of a

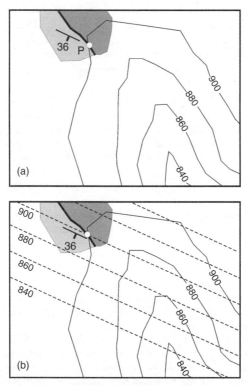

Figure 4.5 *Use of structure contours to interpret a contact. (a) Strike and dip have been measured at point P; (b) contours are drawn parallel to strike through P at a spacing calculated by contour interval/tan (angle of dip); and (c) the predicted outcrop is where the calculated structure contours intersect the topographic contours.*

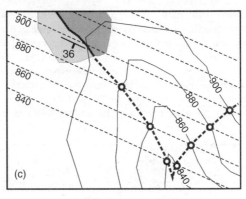

Figure 4.5 *(continued)*

contact at which strike and dip have been measured. From any point P on the mapped contact, a structure contour is drawn for the contact. This contour runs parallel to the measured strike and has an elevation equal to the ground height at P. Parallel contours can then be drawn (Figure 4.5b) with a spacing given by

Contour spacing = contour interval/tan (angle of dip)

Predicted positions of outcrop are given by intersections of structure contours and topographic contours of the same elevation (Figure 4.5c).

4.6 Superficial Deposits
Depending on the nature of the mapping project, superficial deposits can be looked upon either as a nuisance that hides the more interesting solid geology or an important source of information on geological processes that operated during the last 2.6 million years.

In the UK, the maximum southerly extent of the Quaternary Anglian and Devensian ice-sheets lies along a line running roughly from the Severn Estuary to the Thames. North of this line you will expect to find superficial glacial deposits; south of this line there will be no glacial tills present but the rocks may be deeply weathered by freeze-thaw periglacial action.

Unconsolidated superficial deposits are the debris resulting from rock weathering during the formation of the landscape, for example from deposition of sediments by glaciers or river systems. Such deposits include scree (*talus*), which forms unstable slopes of mainly coarse unsorted fragments of broken rock, and 'colluvium', a general term for the rocky hillside soils shed by rock weathering and the poorly developed soils on the lower slopes. They also include the well-developed and thicker soils on lower-lying ground formed largely by rock weathering in place. Except for scree, those need not be shown on your final solid geology map although they appear on your field map as the material that occupies the space between rock outcrops. However, notes can be made, such as 'red soils' to justify the concealed continuation of a dolerite dyke; or 'sandy soils' to reflect underlying acid rocks (see Section 4.6.1).

Alluvium is unconsolidated material transported by flowing water and deposited where the stream slows, and it includes everything from boulders and gravel, through sands, to silt and clay (see Appendix D, Table D.2). Then there is 'drift', which embraces a range of sediments of glacial and fluvio-glacial origin, for example boulder clays, moraine; beach deposits of many grain sizes; wind-blown sands forming dunes (yes, even in Britain) and fine-grained loess. In some countries wind-blown volcanic ashes cover huge areas. More consolidated, but still superficial weathering products include laterites, whilst some spring-fed streams may deposit travertine terraces. All these you should show on your maps.

In general select what is important to the later history of your area and for the principal purpose of the map. If necessary prepare a separate 'drift map'.

4.6.1 Evidence from float

Many soils, particularly on hill slopes, contain rock fragments called *float*. Fragments from the more resistant rocks tend to be large and may lie on the surface. Those from softer rocks are smaller and usually buried; they have to be dug for with the sharp end of your hammer or with an entrenching tool. Contacts on hillsides can sometimes be located with considerable precision by searching for the upper limit of float derived from a formation that lies immediately below a contact with another rock (Figure 4.6). Care must obviously be taken in glaciated regions that the hill slope soils, the colluvium, have not been transported.

4.6.2 Landslides

Landslides can occur on shallow and steep slopes depending on the competence of the underlying rock material, its bedding orientation, jointing pattern and water content. Landslides need to be recognised by the field geologist when geological mapping. Landslides can be small or large (Figure 4.7), active or relict. Active landsides are usually recognisable by the presence of back-scarps, fissures and recently mobilised material. Ancient relict landslides in contrast may not be obvious at all and may have blended into the natural landscape over thousands of years. If the geologist does not recognise that he or she is standing inside a large landslide complex then a great deal of effort could be wasted making observations on material that has slid down-slope and has become

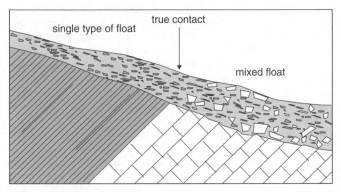

Figure 4.6 *Float in hillslope colluvium as an indicator of a contact; the contact is where the first signs of mixed float appear in the soil.*

Figure 4.7 *Digital infra-red image of the Mynydd yr Eglwys landslide in the upper Rhondda Fawr valley, South Wales, taken in 2005. Important elements of the landslide's morphology are indicated.*

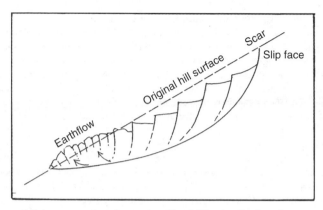

Figure 4.8 *A cross-section of a rotational landslide, showing how the toe extends further out than the original base of the hill slope, resulting in a lower average gradient, and why the surface of the slide consists of small ridges and hummocks with small streams and ponds between them. There are other types of slide too.*

detached from the bedrock. The bedrock mapper may choose just to mark a large landslide zone on the geology map without recording further details and move on. If you are an engineering geologist concerned with hazard mapping or infrastructure development (roads, railways, housing, etc.) you will spend considerable time creating a detailed factual geomorphological map of the landside (Crozier, 1986).

Landslides have many forms such as falls, topples, slides, spreads and flows. In fact most are complex, exhibiting many of these features in one landside (Cruden and Varnes, 1996). Figure 4.8 illustrates the main features of a rotational landslide. Often the most prominent landscape feature is the major arcuate scar or back-scarp, which is the point of origin. Up-slope of this may be more evidence of fissuring. Down-slope of this point lies the material that has slid to its furthest extent at the toe (McDonald, 2000). The parts of the landslide that are in tension can be mapped by the presence of cracks and fissures. The parts of the landslide under compression exhibit folding and display parallel ridges and hummocks. A key observation on moving landslides is that trees that once grew vertically now grow at an angle or are bowed at their base as they strive to grow vertically. Mapping drainage and spring lines is very important as the presence of groundwater increases the internal pore pressures, which then further mobilises the landside.

4.6.3 Pitting, trenching, augering and loaming

When it is essential to examine rock beneath the soil in a poorly exposed area, trial pits and trenches may have to be dug. If the soil cover is very thin it is just a matter of digging away with your geological hammer. If the soil cover is thicker than 0.5 m you will probably need to dig a trial pit using a mechanical digger, and this will cost money. This is common practice in engineering site investigation, where trial pits are regularly dug using a JCB or similar mechanical excavator. Rock samples are excavated by the JCB bucket, from which a geologist then collects a sample from a safe distance away. The walls of the trial pit may be also logged by inspection or photographed from a safe distance outside the pit.

Geologists must not enter into any trial pit greater than 1 m deep; it is too dangerous. Be aware that geologists have died from being trapped inside a collapsing unsecured trial pit. Hand digging deep trial pits in loose sand is highly dangerous. Guidelines exist under British Standard BS5930:1999 *Code of practice for site investigations*, and these should be followed at all times.

4.7 Drilling

Every professional geologist will be concerned with drilling at some stage of his or her career. Drilling technology can range from the simplest petrol drilling hammer systems, to the multi-million-dollar steerable drill rigs used in the oil and gas industries to penetrate depths of many kilometres. Drilling is most commonly employed to prove rock formations at depth, to solve structural problems or to obtain deep samples of rocks and ores. It is used to find minerals, exploit water, in ground engineering and of course hydrocarbons.

Basically, there are two kinds of drill: percussion (or churn) and rotary. Percussion rigs drill by repeatedly raising a heavy drill bit attached to a wire cable and dropping it to strike the bottom of the hole. An alternative method is the use of a vibrocorer, which is an unbalanced mass attached on top of the drill bit, which is then rotated using a portable hand-held petrol motor. In both cases, rock is crushed and chipped away by the drill bit. At intervals, the debris is bailed from the hole with water for examination or simply retrieved by jacking the drill bit back out of the hole.

Rotary drills, on the other hand, rotate a drill attached to the end of a tubular drill pipe or rod: the rock is gradually ground away (Figure 4.9). To reduce the friction on the drill bit water is pumped down the drill stem (water-flush systems); alternatively in softer sediments compressed air is blown down (air-flush systems). Frequently, but by no means always, the bit is set with industrial diamonds, hence the term 'diamond drilling'. Some rotary bits are tubular and cut a ring-like hole, which leaves a cylindrical 'core' of rock attached to the bottom of the hole. This core sample can be broken off and retrieved using a wireline system without the need to bring all the rods back out of the hole (wireline drilling).

Figure 4.9 (a) Wireline, waterflush, rotary, diamond drilling exploration at Dolaucothi gold mine, Mid-Wales. (b) Cores retrieved from drilling. (c) Cores sequentially stored in core boxes and indexed against depth.

Percussion rigs can drill only shallow vertical holes: they yield chippings and rock flour without a core sample from holes about 20–60 cm in diameter. Rotary rigs can drill inclined holes and in the hydrocarbon industries are remotely steerable, even capable of drilling horizontal holes. Rotary drilling may or may not result in a core sample. Holes can be 4–60 cm in diameter, with the larger holes drilled by 'tricone bits' consisting of three conical cog-like cutting wheels. The sludge of ground-up rock flour formed during rotary drilling is pumped back up from the hole by the circulating drilling fluid (usually water) and collected as mud sample material, whether core is taken or not. Geologists who sample and study this material, usually under a microscope, are known as 'mud loggers'.

4.8 Geophysical Aids to Mapping

Geophysics can play an important role in providing very useful information for any mapping programme. Every geologist should at least know about the wide variety of techniques on offer, the basic physical principles of each and what they are likely to be able to detect. The application of most techniques requires a trained geophysicist to apply and interpret the data, but some methods can be used after a few hours of basic training.

Applied geophysical techniques fall into two broad categories – passive and active. Passive techniques just measure already existing natural Earth fields using a receiver. Active techniques are geophysical experiments where the geophysicist is in charge of some kind of active energy source and he or she measures the interaction of that source with the Earth using a transmitter and receiver system.

There are four main passive systems:

- *Gravity measurements:* The gravity at any point on the Earth's surface varies due to the distance away from the centre of the Earth, tidal effects, local terrain and local density variations in the Earth's crust. Using a very expensive gravity meter, very small spatial variations in gravity can be measured. Applying numerical corrections for latitude, height above sea level, tides and local terrain we arrive at a value of Bouguer gravity. Bouguer gravity values relate to local crustal density variations. Negative Bouguer gravity means that the local crustal density is lower than average (e.g. acid igneous rocks or sedimentary basins). Positive Bouguer gravity values are the opposite, indicating that the local crustal rock densities are higher than average (e.g. basic igneous rocks). Bouguer gravity maps of UK areas can be obtained from the BGS.

- *Magnetic measurements:* The Earth has its own magnetic field, but other magnetic fields can be produced by subsurface rocks containing the mineral magnetite. By mapping the Earth's field using a portable magnetometer the presence of magnetite can be detected in buried geology by mapping local anomalies in the overall background Earth's field. Magnetometry is popular in archaeological mapping but is also regularly used in field geological mapping to map basic igneous dykes and sills or metalliferous orebodies containing magnetite (Figure 4.10).

- *Electrical self-potential:* If the local geology contains metalliferous minerals below the groundwater table, electrochemical reactions can produce a natural electrical voltage rather like a car battery. These voltages, typically less than 1 V, can be mapped using a simple voltmeter and two non-polarising electrodes.

- *Radiometry:* Acid igneous rocks, rich in potassium feldspar, contain sufficient ^{40}K (potassium-40) to enable them to be distinguished from rocks with less K-feldspar nearby if a sufficiently sensitive instrument is used and the soil cover is thin. A gamma-ray spectrometer (scintillometer) will detect these differences although the older *Geiger counter* cannot.

Figure 4.10 *Proton magnetometer survey being carried out to detect and map a buried gabbroic dyke at Rhoscolyn, Anglesey, North Wales.*

The more sophisticated active systems include:

- *Seismic refraction:* The travel times and velocities of compressional P-waves and shear S-waves are measured through the shallow subsurface. Usually a sledgehammer acts as the seismic source and a recording system of 48 geophones is used to measure the travel times. By interpreting the travel times and ray paths taken through different geological layers, simple estimates of the depth to bedrock under the superficial cover can be made.
- *Seismic reflection:* Seismic reflection is the main technique employed in the multi-million-dollar hydrocarbon industries on land and sea. 'Images' of the subsurface in 2D or 3D are produced by bouncing compressional P-waves off each geological layer in turn down to many kilometres depth. The reflection method is highly complex and financially beyond most mapping exercises.
- *Electrical resistivity:* 2D or 3D 'images' of the electrical resistivity properties of the subsurface are made by driving electrical currents through the ground using a 12 V car battery and a pair of steel electrodes. Voltage measurements

are made between another pair of electrodes. Modern systems employ multi-electrode cables capable of automatically collecting hundreds of readings to make an 'image'. Resistivity can be used to find the faulted contact between two rock units of different resistivities or the presence of metallifeous ores, but a popular application of electrical resistivity is for hydrogeological exploration. The presence of groundwater makes a rock electrically conductive, so shallow zones of high conductivity within an 'image' can imply the presence of groundwater.

- *Electromagnetic ground conductivity:* Electromagnetic (E/M) waves are sent from a transmitter to a receiver unit. As well as a primary wave travelling though the air to the receiver, the E/M waves also penetrate some distance into the subsurface geology. If the ground is electrically conductive, the E/M wave produces a secondary E/M wave, which is also picked up by the receiver. Using complex electronics, the secondary wave can be compared to the primary wave to arrive at a value for the average ground conductivity beneath the instrument. The depth of investigation can be controlled by changing the frequency of the E/M signal and the spacing between the transmitter and receiver. 'Metal detectors' are one type of ground conductivity instrument, but for geological mapping more specialised systems with greater depths of penetration are employed. Systems such as the EM-31 or GEM-2 are often linked up to GPS units and set to automatically take a ground conductivity reading every second together with its location. By walking over a mapping area with such a system, a ground conductivity map can be made as an aid to geological mapping (Figure 4.11).

- *Ground probing radar (GPR):* This is another type of sophisticated E/M system that bounces pulses of high-frequency E/M radar waves off subsurface geology to produce an 'image'. However, in normal geology with a thin superficial cover, the depth of penetration is very limited. Although ground probing radar (GPR) is widely used in archaeological mapping it has limited use in geological mapping.

Of the nine geophysical methods briefly described above, the three that are most commonly used regularly in combination with geological mapping are magnetometry, radiometry and E/M ground conductivity. All three can also be adapted and deployed from aircraft or helicopters for reconnaissance surveys of large areas. They all can be used to create geophysical maps that can be overlain on a geological map, especially using GIS systems, to fill in knowledge gaps where rock exposure is poor. Potentially all the other geophysical methods can be employed to solve localised specific geological problems within a mapping programme.

Figure 4.11 *GEM-2 electromagnetic instrument coupled with a dGPS position-ing system being used to help map faulted geological boundaries in the Rhondda Valley, South Wales.*

4.9 Large-Scale Maps of Limited Areas

From time to time there is a need to map specific aspects of the geology on a far larger scale than that used for your main map. You may be able to pho-tographically enlarge part of your base map or, in Britain, use a 1:2500 map. More often, the need arises for a very large-scale sketch map of a very limited area, sometimes only a few 100 m in extent. The need is to illustrate geology and no great precision is required. Thus methods can be used that might well be derided by a land surveyor. Some are described below; they can be modified or changed to meet contingencies. Ingenuity and a basic knowledge of surveying are assets. Keep sheets of squared paper in your map case should you need them.

4.9.1 Compass-and-tape traverse

The simplest method of plotting geological detail is by taking offsets from a chain line or traverse, as described in Section 3.5.4. A single traverse may even suffice (Figure 4.12). The same method can be used as a 'mini traverse' to map a single large exposure in detail.

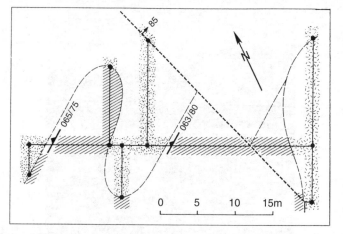

Figure 4.12 *A compass-and-tape traverse to plot larger-scale geological features.*

4.9.2 Traverse with offsets

Where a number of exposures are spread over an area of more or less level ground, but scattered too far apart to be mapped by a single traverse line, geology can be rapidly mapped by running a series of traverse legs in a loop, ending at the start. Detail is mapped by offsets from the traverse legs (Figure 4.13). For small areas, measure the traverse legs first, marking the turning points on the ground so that they can be easily found again. Plot the traverses and correct the closure error (see Appendix A), then plot the geological detail. The alternative is to enter all details, including the geology, in your notebook as you move along each leg in turn, and re-plot everything back in the camp. The first is to be preferred because then you have the ground in front of you as you plot in the detail.

4.9.3 Mapping an exposure in detail

It is sometimes necessary to map a large exposure in detail. If the surface is more or less flat, lay down a base line; use stones to mark points along it at fixed intervals (say 10 m); then measure traverses at right angles from it, with stones again at 10 m intervals. The effect is to build up a grid to guide your sketch map (Figure 4.14). Where a great deal of sketch mapping is to be done, a cord grid that can be laid over an exposure and anchored there with stones will simplify the task. The grid shown in Figure 4.15 was constructed by first

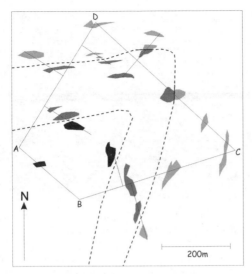

Figure 4.13 A closed traverse of several legs to plot in a number of exposures for a medium-scale sketch map.

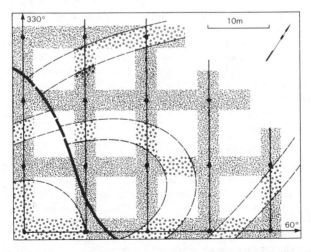

Figure 4.14 Mapping a large exposed area by building up a rough grid.

Figure 4.15 *A 4 m cord mesh, used as a grid to aid sketch mapping, laid over an exposure at Badcall, NW Scotland (see Figure 4.16).*

pegging out an area of 16×20 m (on the beach) with pegs every 4 m along the sides. Three-ply nylon cord was used to make a net with a 4 m mesh. Detail is plotted by estimation on squared water-resistant paper, with measurements using a steel pocket tape where necessary. Compass bearings are measured by assuming one side of the grid to be 'grid north' and correcting your compass to read accordingly. Figure 4.16 shows structure mapped in the deformed 'Scourie' dyke shown in Figure 4.15.

Remember that other methods not described here may be used by other people. Choose the methods that will suit you and the geology best, but ensure that those you do use give acceptable results in terms of the accuracy required. Different types of geological environment will affect the way you map, so will different terrains, different climates and different base maps. Adapt and invent.

4.10 Underground Mapping

Although mapping geology in underground workings, especially in metal mines or in caves, is not a subject for this small book, some very basic principles can be described. Exploration geologists wanting to learn how to carry out underground mapping in detail should consult Forrester (1946), for his excellent examples of coloured underground field sheets, and also Peters (1978) and Marjoribanks (2010). The difference between surface and underground mapping is that in underground mapping you are actually inside the geology in 3D space, working

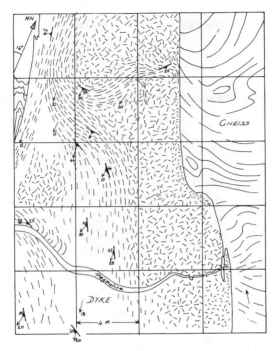

Figure 4.16 *Structures mapped in a deformed 'Scourie' dyke at Badcall, mapped using the 4 m cord mesh shown in Figure 4.15 (courtesy of R.H. Graham).*

in tunnels at varying depths and changing azimuths (Figure 4.17). You cannot map the tunnel floor as it is coved in debris and mud, but you can map both tunnel walls and possibly also the roof. Conventionally therefore the geology in metal ore mines is normally mapped from the walls at waist height or projected to waist height. It is essential that a mine is surveyed in detail first, as without that knowledge of where you are underground any geological mapping cannot be directly related to a surface map. By consulting any available mine plans from within your study area, very useful additional subsurface geological information can be obtained.

It is often a temptation for geologists to enter abandoned mine workings or caves in an area they are mapping. Resist it for reasons of safety, unless part of an official mine visit made under supervision (see Section 1.2).

Figure 4.17 *Students undertaking an underground geological mapping exercise at Dolaucothi gold mine, Mid-Wales. Note the tape measure laid out along the adit floor marking distances away from a mine survey point. Mine survey points are usually marked by bolts drilled into the adit roof.*

4.11 Photogeology

Photogeology is the systematic interpretation of geology from aerial photographs. It can be used as a method of geological reconnaissance with only limited ground checking, or as an adjunct to orthodox geological mapping. Here, we consider only its second use.

4.11.1 Using aerial photographs

Before leaving for the field area, examine your photographs under a mirror stereoscope and make an interpretation of the main geological features. Do not take the original images into the field; make copies and laminate them to make them rainproof. For the field you can either use conventional stereo techniques with a pocket stereoscope or photographically combine the images into an anaglyph image that can be viewed using 3D (red/cyan) glasses (Section 2.12).

When you reach the field, carry the photographs in your map case in addition to your field map. Examine the photos at intervals to compare what you see on the ground with its appearance on the photographs. Back at camp, or in the evening, review your map and stereo photographs again. The vertical exaggeration of the 3D image often accentuates quite minor landscape features which reflect geology. Check in the field next day to see if you can now locate these features on the ground.

Also examine on the ground any other interesting features you have seen on the photographs whose geological cause was not obvious; their geological significance may now become apparent. Often photographs will point you towards places on the ground you might otherwise not have bothered to visit. Some indications on photographs, however, you may never be able to solve. This does not mean that they do not exist; show them on your map and fair copy maps in purple so that future workers are aware of them. Eventually their significance may be found. Remember that photogeological evidence is not inferior to other geological evidence; it is merely different.

4.11.2 Photogeological features

Only a few indications of what can be inferred from photographs can be given here. Refer to Ray (1969) and Lillesand *et al.* (2008) for further information, but experience is the best teacher. Note the following advice.

- *Tone* results from ground reflectivity. It varies with changing light conditions. Sudden changes of tone on a single photograph may indicate a change in rock-type owing to a change in vegetation or weathering characteristics.
- *Texture* is a coarser feature caused by erosional characteristics. Limestones have a rough texture; soft shales are often recognisable by a micro-drainage pattern.
- *Lineaments* are any straight, arcuate or regularly sinuous features of geologically uncertain significance seen on photographs. They may show in the drainage as vegetation changes: thin lines of lusher vegetation in arid bushland, perhaps resulting from faults, master joints, contacts or for some other geological reason allowing water to seep closer to the surface. The cause of some lineaments may never be discovered.
- *Vegetation* is an excellent guide to geology and changes can usually be more easily seen on colour photographs than on the ground. It contributes to both tone and texture.
- *Alluvium, swamps, marshes* and so on are quite distinctive on photographs and their boundaries can usually be mapped better from photographs than on the ground.
- *Strikes and dips* can be seen from dip slopes, scarp edges and from the way in which the beds 'vee' in valleys. There are even methods and instruments for calculating the amount of dip where large dip slopes are exposed.

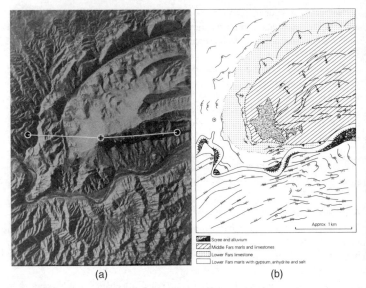

(a)   (b)

Figure 4.18 (a) Aerial photograph from Iran compared with its photogeological interpretation, and (b) note that the symbols used are different from those used on ordinary geological maps.

4.11.3 Systematic analysis

Only a brief description of systematic photogeological analysis can be given here.

1. Tape an overlay of plastic drawing film (Permatrace, Mylar or similar) over one photograph of a stereopair (Figure 4.18a); mark on the *pp* and *cjs* (see Section 3.7)

2. Under the stereoscope, trace the drainage onto the overlay (in black) to provide a topographic framework. Include alluvium and terrace boundaries. Outline the areas of scree, landslide, outwash, and so on.

3. Trace (in purple) scarp edges and indicate the direction of dip by arrows down the dip slopes; the steeper the dip, the more barbs on the arrows (Figure 4.18b).

4. Draw, again in purple, any known marker beds that can be traced. Indicate the dip by 'ticks' (Figure 4.18b): the steeper the bed, the greater the number of ticks.

5. Show obvious faults in red.

6. Plot as lineaments all major linear and arcuate features whose cause is uncertain. Show them as purple lines, broken at intervals with three dots.
7. Draw contacts as dotted lines, again in purple.
8. Identify rocks and label formations.

Check your interpretation on the ground and against your field map. Amend as necessary and transfer your photogeological information to your field map in the appropriate colours to distinguish photogeological data from other information. If you are mapping directly onto transparent overlays to photographs in lieu of a field map, show any information mapped or confirmed on the ground in black. Always distinguish the two sources of information. After the mapping programme is complete your interpreted photogeological overlay can be digitally scanned and geo-referenced to form a layer in a GIS database.

5

TECHNOLOGICAL AIDS TO MAPPING

Geological mapping involves the mastery of a wide range of skills: observational and interpretive skills, a broad knowledge of rocks and geological processes, plus navigational and cartographic skills. This chapter will concentrate on the latter two, which have seen significant changes in recent years due to major technological advances. The equipment and data resulting from these advances help geologists produce very accurate 3D topographic base maps, and achieve improved locational accuracy whilst in the field; and geologists now have the potential to visualise their final map in 3D.

5.1 Digital Terrain Models

In Section 3.7 we discussed the way in which geologists use aerial photography as an aid to mapping and how for over half a century stereo pairs of photos have been used to create topographic maps. Over the past decade new ways of creating detailed and accurate digital topographic models of the Earth's surface have been developed, and these are proving a great assistance to geological mapping. These computer landscape models are known as Digital Terrain Models (DTMs) or Digital Elevation Models (DEMs); they are mathematical approximations of the complex 3D topographic surface of a given study area. These DTMs are creating a revolution in the way we look at the Earth's surface. Terrain models are used by geologists to:

- Create a detailed landscape 'picture' of the study area as an aid to reconnaissance prior to the field programme. Terrain analysis techniques can also, for example, be used to remotely calculate topographic cross-sections or slope angle maps.
- Produce an accurate base map for field mapping.
- Help create 3D landscape visualisations. The final geology map can be draped onto the 3D landscape model to display the relationships between geological structures, lithologies and landscape morphology.
- Assist in global tectonic/isostatic studies, by measuring large-scale land movements as a consequence of regional tectonic uplift or earthquakes.

Basic Geological Mapping, Fifth Edition.
Richard J. Lisle, Peter J. Brabham and John W. Barnes.
© 2011 John Wiley & Sons, Ltd. Published 2011 by John Wiley & Sons, Ltd.

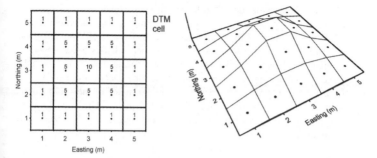

Figure 5.1 *Visual representation of a Digital Terrain Model (DTM). The landscape is divided up into square cells and the average elevation of each cell calculated.*

5.1.1 Basic principles of DTMs

To create a DTM of any area of the Earth's surface, the area is divided into square survey cells and the average topographic elevation of each cell (above a datum) is determined (Figure 5.1). The quality of any DTM depends on the size of the survey cell and the precision of the average elevation value. In computer terms, a DTM in its simplest form is just a list of millions of x, y and z data values, with x and y being the geographical coordinates of the centre of the cell and z its elevation value. The footprint size of a DTM cell may range from 25×25 m for regional reconnaissance surveys, to 1×1 m for detailed applications. For geological purposes the 'bare Earth' DTM model is preferred, where the cell elevation value is the average ground level stripped of vegetation and man-made structures.

5.1.2 Methods of visualising DTM data

A DTM containing millions of individual cell values requires dedicated GIS software to process such large amounts of data. Software such as ArcGIS, Mapinfo, Surfer, Global Mapper and so on are all able to create landscape visualisations from DTM data files. Such datasets can be visualised in a number of ways (Figure 5.2):

- *Image maps* – The DTM is viewed as a digital raster image from vertically above. The colour or greytone assigned to each individual cell is controlled by the z (elevation) value of that cell.
- *Shaded relief maps* – A 3D surface model is created from the DTM, which is normally viewed as a digital raster image from vertically above. Artificial

81

sunlight is shone across the 3D surface from a point source 'sun'; the position of the 'sun' can be moved by the software operator to any compass direction and vertical azimuth. This creates light and shaded areas for ground slopes facing towards or away from the 'sun'. Shaded relief is a very powerful technique used by geologists to reveal the intricate details of a complex landscape and also to accentuate subtle landscape lineaments.

- *3D surfaces* – The software creates a solid 3D model of the DTM surface by interpolating between the individual DTM points. This model can be viewed from any perspective angle, vertically magnified to any degree or preferentially illuminated as in the shaded relief description above.

5.1.3 Techniques for acquiring digital terrain data

Detailed and accurate DTMs of any area of the Earth can be acquired by:

- Digitising topographic contours from paper-based maps.
- Photogrammetric analysis of stereoscopic aerial photos or satellite images.
- Measuring the time it takes radar reflections to return from the Earth's surface, acquired from aircraft surveys. The huge advantage with radar data is

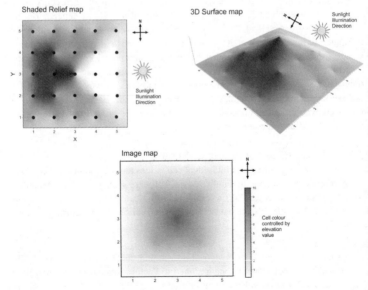

Figure 5.2 *Three ways in which a Digital Terrain Model made up of a number of individual cell values (black dots) can be visualised using dedicated software.*

that they can be recorded either by using short-wavelength pulses as first reflections from vegetation digital surface model (DSM) or by using long-wavelength pulses from the underlying ground surface (DTM). This allows geologists to 'see through' thick vegetation cover to reveal bedrock outcrops. In tropical regions of permanent cloud and thick vegetation, radar-derived DTMs are revolutionising geological mapping.

- Using lasers (Light Detection and Ranging, LIDAR) fired from scanners housed on aircraft. As with radar above, the time it takes laser reflections to travel from a moving aircraft to the Earth's surface and back can be converted into distance measurements. By constantly monitoring the aircraft's position with GPS and using ground control points, a very detailed terrain model can be constructed. Laser reflections will be from the first surface encountered (vegetation or man-made structures), thus producing a DSM.

- Using differential GPS, Global Satellite Navigation Surveying System (GNSS) or Real-Time Kinematic (RTK) surveying techniques (see Section 5.2).

Details of all these techniques can be found in specialised textbooks (e.g. Heritage and Large, 2009; Liu and Mason, 2009; Vosselman and Maas, 2010; Wilson and Gallant, 2000). Here we give some examples of DTM data derived by the above methods and discuss their applications for geological mapping (Figures 5.3 and 5.4).

5.1.4 Terrain analysis techniques

In the previous section we have treated digital terrain images as either a way of obtaining accurate landscape 'pictures' to aid geological interpretation or to use as base maps. However, other techniques for the analysis of terrain models can be used to assist mapping:

- *Slope angle analysis* – The first derivative of the 3D landscape can be calculated to produce a map of ground slope angles from the DTM. This is particularly useful in landslide mapping, highlighting areas of steep slopes. The second derivative of the 3D landscape surface can be calculated to highlight rapid changes in slope angles. These 'breaks-of-slope' can be used to track geological outcrop boundaries and fault lines (see Section 4.5.3).

- *Slope aspect models* – The compass direction in which a slope faces is called its aspect. Maps of slope aspect can be calculated from the DTM; these are of particular use in mapping periglacial freeze-thaw regions and snow avalanches. Slopes that face towards the sun are more prone to melt during the summer months.

- *Derivation of drainage systems* – Using the 3D landscape model and the principle that rainwater falling on a surface always flows down slope, 'virtual' precipitation can be modelled onto the DTM. Such computer modelling can then reveal the predicted patterns of stream or river channels that develop on the landscape, and determine individual river catchments, drainage basins and floodplain areas at risk of flooding.

5.1.5 The future of terrain models in geological mapping

Airborne LIDAR technology has rapidly developed for the purpose of topographic mapping. In England and Wales, the Geomatics Group of the Environment Agency has been systematically mapping with LIDAR since

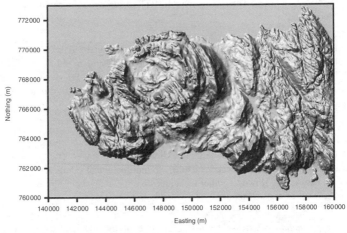

Figure 5.3 *A 20 × 13 km DTM of the Ardnamurchan Peninsula in NW Scotland, UK, displayed in shaded relief with an artificial illumination from the west. The ring-like topographic features are a consequence of a Tertiary igneous complex surrounding a central caldera. DTM data are derived from Ordnance Survey contour mapping (profile data) compiled using 2.6 million 10 × 10 m individual data cells at 1.5 m elevation precision. Being contour-derived elevation information, there is an element of smoothing already applied to the landscape model. Such DTM data are just sufficient in resolution for geological mapping in areas of high topographic relief. However, the 1.5 m vertical precision is not usually detailed enough to create accurate geological base maps in areas of low topographic relief.*

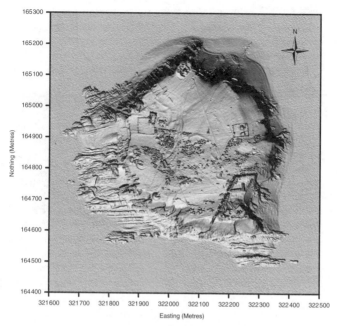

Figure 5.4 *A 900 × 900 m LIDAR-derived Digital Surface Model (DSM) of Flatholm Island in the Bristol Channel, UK, captured at low tide. The DTM is displayed in shaded relief with an artificial illumination from the south. The uplifted Flatholm Island comprises Carboniferous limestone deformed by Variscan folding and thrusting. The island has a very thin soil cover. The LIDAR data are supplied by the UK Environment Agency (EA) as part of their ongoing UK mapping programme. LIDAR data are supplied at a 2 × 2 m ground cell resolution, with a vertical precision of around 15 cm. The image contains 0.2 million data points. Such high-quality LIDAR data can produce an excellent base map for geological mapping. For large study areas covering many square kilometres, LIDAR data files can be excessively large and expensive to obtain, even at educational rates. Copyright: Permission obtained from UK Environment Agency, 2011.*

1997, achieving 60% land coverage by 2010. These EA data are available at reduced educational rates for geological mapping exercises. In the UK, free LIDAR data are available for educational purposes through the Mimas Landmap project, and also for southern UK coastal areas from the Channel Coast Observatory website.

Figure 5.5 *Terrestrial laser scanner being used by Terradat UK Ltd to make 3D scans of the geological cliff section at Nash Point, South Wales. Repeated scans are taken at regular time intervals and compared, to quantify rates of cliff erosion. Photo: Nick Russill, Terradat Ltd. Permission obtained from Terradat, UK.*

Airborne-based LIDAR cannot map vertical cliff sections. Geologists are now experimenting with ground-based (terrestrial) laser scanners as a way of mapping rock outcrops and coastal cliff sections (McCaffrey *et al.*, 2003, 2005). This allows the detailed mapping of structures such as exposed fault surfaces or folded bedding surfaces, or quantifying rates of erosion by the repeat surveying of topographic surfaces (Heritage and Large, 2009; Figure 5.5).

5.2 Topographic Surveying Techniques

The geologist requires a basic knowledge of surveying to construct an accurate scale base map, topographic cross-section or even a topographic contour map of a small region within an overall study area. For example, you may also need to work out the accurate dimensions of a cliff face or rock outcrop, or to calculate the volume of rock extracted from a quarry. In these cases, a simple compass, tape measure and autonomous GPS cannot produce the required levels of surveying accuracy, particularly when measuring vertical relief. In the following sections, the basic techniques and equipment used for more accurate surveys are briefly described. This is a textbook for field geologists not surveyors, so only the very basic surveying principles are covered here. For further details, students are directed to modern textbooks on practical surveying techniques (e.g. Uren and Price, 2010). There is an ever-increasing diversity of surveying equipment available. Up-to-date technical information can be obtained from the websites of equipment manufacturers such as Topcon, Trimble, Leica, Nikon, Pentax and Sokkia.

5.2.1 Optical levelling techniques

Optical levelling is a simple technique for obtaining relative topographic height measurements of a number of individual survey points. An optical level (or dumpy level) is basically a horizontal telescope mounted on a tripod with a cross-hair in the optical viewer (Figure 5.6). The instrument, once secured to a tripod, needs to be levelled. The telescope then will rotate in the horizontal plane through 360° allowing the operator to take accurate relative horizontal bearings (note there is no reference to magnetic north). A second person is required to hold a 3 m long vertical staff (graduated at centimetre intervals with zero at its base) at the ground survey points (Figure 5.7). This person should take great care whilst working around overhead power lines.

To carry out a survey, the operator looks through the level and focuses on the staff being held vertically at a survey location (Figure 5.8). Using the central cross-hair, the height above the ground is read off the staff and noted. This is the *backsight*. The person holding the staff then moves to a second location, B, the level is rotated in the horizontal plane, the above process is repeated and the *foresight* value is noted down. The height difference (H) between the two points is the backsight minus the foresight reading (take care to note down the sign, plus or minus, in this calculation, as this tells you if B is topographically above or below A). The horizontal distances X and Y are usually measured by using

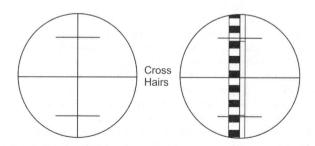

Figure 5.6 *View through the eyepiece of an optical level, showing a typical cross-hair system and centimetre graduated staff.*

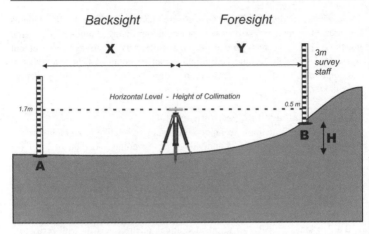

Figure 5.7 Principle of the levelling technique. See text for details.

a tape measure laid out along the ground. To construct a long survey traverse, successive foresight readings are taken until the staff is out of view, that is, the view through the eyepiece of the central cross-hair is projected below the bottom, or above the top of the 3 m staff. In this case the tripod and level have now to move to a new survey point and the process is repeated. The fundamental rule is that either the staff or the tripod moves between successive survey points, never both. Treat this as a new survey, and the two separate surveys are linked together by reoccupying common survey stations. After completing a topographic cross-section from beginning to end, it is recommended to repeat the survey, from the end back to the beginning, to double check your results. This is known as closing the survey (Appendix A). If you have done everything correctly the closure error should in theory be zero; however, around 10 cm of closure error is usually an acceptable precision for a geological survey. If the closure error is in the order of metres then you need to go and re-check your calculations, or in the worst case repeat the survey.

Note that when measuring steep topographic slopes the tape measure laid out on the ground will measure the slope distance between the two survey points, not the horizontal distance, which you need to plot on a map. However, as you have calculated H and know the slope distance, you know two sides of a right-angle triangle, and using Pythagoras' theorem you can easily calculate the corrected horizontal distance. Ideally, one of the survey points you measure should be a benchmark, spot height or trig point with an identified height above

Figure 5.8 *A levelling survey in Cardiff Bay.*

sea level; otherwise all your height measurements will be only relative to an arbitrary height datum.

Levelling is primarily used by geologists to make topographic cross-sections along a traverse by making numerous height and distance measurements. When making a cross-section, do not take measurements robotically at set distance intervals, but survey in breaks-of-slopes to get an accurate reflection of the changing landscape. It is easy to make simple numerical mistakes, so it is recommended to do all the numerical calculations in the field where you are more likely to spot your mistakes and be able to correct them straight away.

5.2.2 Total stations

To overcome the practical difficulties of levelling in areas of rugged relief, a total station can be used. A total station is a surveying instrument consisting of a horizontally mounted telescope that is free to rotate not only in the horizontal but also in the vertical plane (Figure 5.9).

A total station measures the angle of the telescope relative to the horizontal (Figure 5.10). The instrument also has an in-built electronic distance meter (EDM), which measures the travel time of an emitted pulse of low-power infra-red laser energy that is reflected by a prism reflector mounted on an adjustable staff held by a second person at a survey point (point B in Figure 5.10). Using

Figure 5.9 A total station in operation.

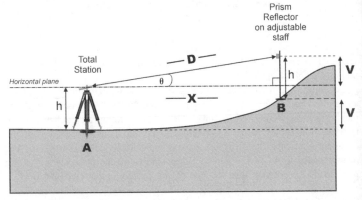

Figure 5.10 The principle of making measurements with a total station. See text for details.

an average value for the speed of light in air, laser reflection travel time can be converted into a distance with millimetre precision.

To take a measurement with a total station, firstly a survey point (A in Figure 5.10) is sited at a place with good visibility. The total station is secured onto a tripod and levelled using a bubble level. The height of the prism reflector mounted on the staff is adjusted so that its height (h in Figure 5.10) exactly equals the height of the work station's measuring level above the ground. The second operator then stands at the survey point B and holds vertically the staff with the mounted prism. The total station is rotated in both the horizontal and vertical plane by the operator until the cross-hairs are aimed at the prism reflector; the optics are locked in position to stop any further movement, then the EDM is fired. Values for the angle θ and distance D (Figure 5.10) are calculated by the instrument. The instrument uses simple trigonometry to convert these two values into the horizontal distance X and the vertical height V, which are required for map making (Figure 5.10). These calculations are simplified because the height of the total station and the prism above ground level are equal. The operator with the reflector moves on to a new survey point and the process is repeated. For optimal productivity, you should try to take as many measurements as possible by initially setting up the total station at the best vantage point. If you cannot see all the survey points from one vantage point, then the total station will have to be moved to a new location. If so, treat this as a separate survey and join the two surveys together by re-occupying common survey points.

Total stations are widely used by geologists to make rapid measurements especially in areas of rugged terrain, for example of survey pegs secured into moving landslides. In hazardous or inaccessible areas (e.g. volcanoes, landslides and cliffs) more expensive reflectorless total stations (which use high-power lasers that do not require a prism to reflect off the target) are used. In good weather conditions, total stations can make millimetre-accurate measurements over distances greater than 500 m between stations. For more details of total station techniques, students are directed to Uren and Price (2010).

5.2.3 Global satellite navigation surveying systems (GNSS)

In Section 2.11, the use of basic autonomous GPS systems for geological mapping was discussed. With such systems, users can only hope to locate themselves on the ground to around an average precision of $15-10$ m, but this locational precision depends on many factors, is unpredictable and time-variable. Improved locational accuracy can be achieved by not only using the 24 US GPS satellites, but also the 18 Russian GLONASS (Global Navigation Satellite System) and Europe's Galileo satellites (which will be fully operational by 2014). Such systems are known as high-precision GNSS receivers and can process 72 satellite channels (Uren and Price, 2010). Using all the above channels and more sophisticated signal processing, the horizontal locational precision of GNSS systems improves dramatically over autonomous GPS, to around 10 cm.

There are, however, two fundamental physical problems still to resolve. Firstly, GPS satellites show slight variations in their orbit path due to variations in the Earth's gravitational field. These errors can be later allowed for by post-processing the GPS data using satellite orbital correction data available over the internet. Secondly, the troposphere layer of the Earth's atmosphere is up to 13 km thick and contains our weather. The speed of light through this layer varies slightly due to factors such as humidity. This causes local timing variations in GPS signals from the satellites to the ground and results in locational imprecision.

Specialist GNSS receivers are available for GIS mapping. Some also have built-in digital cameras and EDM distance measuring capability. Using wireless Bluetooth technology, the unit can also automatically log many other kinds of geological information. These units allow professional geologists rapidly to capture field data and even automatically download them back to the office via a mobile phone. Back at the office, the data can be input into GIS software for automatic plotting, potentially saving many days of old-style note taking and hand plotting.

5.2.4 Differential GPS systems (dGPS)

The 15−10 m variable locational accuracy using autonomous GPS may be acceptable for outdoor walking, but it is not precise enough for detailed large-scale mapping, for coastal navigation in boats or for the aviation industry. The GNSS systems used by professional surveyors are very expensive, but there is

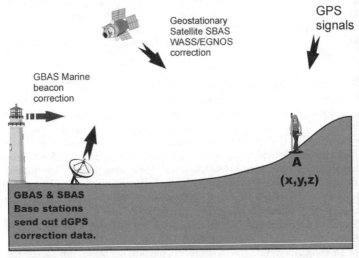

Figure 5.11 *The principle of differential GPS (dGPS) systems. The use of both of Ground-Based Augmentation Systems (GBAS) and Satellite-based Augmentation Systems (SBAS) are illustrated.*

Figure 5.12 *Adapted Trimble marine dGPS system being used for landslide mapping in South Wales. Note marine dGPS antenna housed in rucksack.*

a third, cheaper navigation option for geological mapping. Differential global positioning system (dGPS) techniques have been developed to improve locational precision at a reasonable cost. The principle of dGPS is that permanent ground base stations with autonomous GPS units are located at very accurately surveyed locations on the Earth's surface. The base station tracks all the GPS satellite signals it can see in the sky and works out its position (Figure 5.11). The base unit actually knows its position to centimetre precision; hence, knowing this position and the satellite location, it calculates what the signal timings from each satellite should actually be. The differences between the actual measured timings and the theoretical predicted signal timings from each satellite are calculated; this is the current 'error correction'. This 'error correction' is then transmitted every second as a separate signal to the roving receiver, which applies the same correction to greatly improve its measured location. The further away the rover is from the base station, the more inaccurate the correction becomes, as the corrections are not the same for both instruments. However, dGPS systems can provide around a 1 m horizontal ground precision, which is adequate for most geological mapping surveys (Figure 5.12). For an example of the use of dGPS for detailed mapping of exposed geological structures see Pearce *et al.* (2006).

Ground-based Augmentation Systems (GBASs) and Satellite-based Augmentation Systems (SBASs) exist such as the European Geostationary Navigation Overlay Service (EGNOS) that send out continual dGPS correction information. For maritime applications around the world, various agencies have set up marine dGPS beacons, usually in lighthouses located around the coast. Such systems have still proven accurate for geologists working 100 miles inland of the UK's coastline.

To achieve the ultimate in precision using GPS systems, geologists, surveyors and civil engineers are now using RTK systems in the field. These systems use two GNSS units, a base and a rover. Both GNSS units lock onto many satellites, but also communicate between themselves via a radio link. The semi-permanent base station knows its position on the Earth's surface to around a 10 cm accuracy, but the rover knows its position relative to the base to millimetre accuracy. By just walking around with the rover unit, these RTK systems can rapidly produce millimetre accurate topographic data that can be used to create very accurate cross-sections or DTMs of outcrops for geological mapping.

6

FIELD MEASUREMENTS AND TECHNIQUES

One object of geological mapping is to elucidate the structure and structural history of the region studied. This can only be done if measurements are made of: the attitude of planar structures such as bedding and foliation; linear features including the intersection of bedding and cleavage; the plunges of minor folds and the directions of overturning. It is assumed that the reader already knows what these structures are, although many budding geologists do not know the best way of measuring them. Measurements once made must be plotted and recorded, and there are several ways of doing this too, some easier than others. Structures must also be investigated, specimens collected, photographs taken, and possibly even soils panned to determine heavy mineral suites where no rocks are exposed (see Section 4.6.3). These are all part of the technique of mapping.

6.1 Measuring Strike and Dip of Planar Structures

Measurements of strike and dip of bedding, foliation and jointing are fundamental. Without them, a geological map means little. A useful rule of thumb is to take readings with a frequency of about one per 5 cm^2 of map area regardless of the scale of mapping. Naturally there will be fewer readings in poorly exposed areas, but try to obtain a uniform distribution of data. Do not take measurements only when the strikes or dips have changed!

Strikes and dips can be measured in a number of different ways. Suit your method to the type of exposure. Limestones, for instance, often have uneven bedding surfaces, and a method that allows you to measure strike and dip over a wide area of surface will give more representative values than one where only a point on the surface is measured. Metamorphic rocks offer additional problems. Measurements of cleavage or other foliations often have to be made on very small parts of a surface, sometimes even overhanging ones. There may even be more than one foliation and at least one of them may be obscure and difficult to measure. You must use your ingenuity. Many granite gneisses crop out as pavements or turtlebacks where the trace of the foliation is clear enough but the dip is difficult to see. One point must be emphasised: you must plot measurements onto your map immediately after you have taken them so that any

Basic Geological Mapping, Fifth Edition.
Richard J. Lisle, Peter J. Brabham and John W. Barnes.
© 2011 John Wiley & Sons, Ltd. Published 2011 by John Wiley & Sons, Ltd.

mistakes made in reading your compass – and they do happen – are obvious. This is not the only reason for plotting data directly; the readings on the map define the structure and greatly assist with chasing contacts and the completion of the map. Only in very bad weather is it permissible to log readings in your notebook and plot them back in camp. Joints are an exception. They tend to clutter a map without adding to a direct understanding of the structure. Record joint directions in your notebook and plot them onto map overlays later, or treat them statistically. Another exception to the rule of the immediate plotting of structural measurements is where structures are locally complex: then you may have to draw an enlarged sketch in your notebook and plot the measurements on it. Several different methods of measuring strike and dip are described below; modify them as occasion demands.

6.1.1 Method 1

This, the *contact method*, is commonest of all. Use it where the surface is smooth and even. If there are small irregularities, lay your map case on the rock surface and make your measurements on that, but sometimes such a small area of bedding (or cleavage, fault surface, etc.) is exposed that direct contact is the only method than can be used. Place the long edge of your compass on the surface, hold it horizontally, align it parallel to strike and read the bearing (Figure 6.1). Some compasses are provided with a level bubble so

Figure 6.1 *Measuring the strike by the contact method.*

Figure 6.2 *Using a 'Dr Dollar'-type clinometer to measure dip.*

that there is no difficulty in establishing the strike. With others, you may first have to determine strike with your clinometer by measuring the tilt of different lines in the bedding plane until the untilted strike line is found. Once found, mark the strike line with a scratch of a hammer point or by laying your scale down beside it. Measure dip with your clinometer at right angles to the strike (Figure 6.2).

With practice you can usually estimate strike and dip with sufficient accuracy, but where surfaces are close to horizontal, strike may be difficult to estimate. Then it may be easier to determine the dip direction (direction of maximum dip) or if you have water to spare let a little run over the surface to determine the dip direction. One of us (RJL) has found a plywood board with a mounted spirit-level to be a helpful aid for the contact method (Figure 6.3).

6.1.2   Method 2

On large uneven surfaces of relatively low dip, estimate a strike line of a metre or more in length (if necessary, mark it with a couple of pebbles), then stand over it with your compass opened out and held parallel with it at waist height (Figure 6.4). In a stream or on a lake shore nature may help, for the water line makes an excellent strike line to measure. The same method can be used to measure the strike of foliation or of veins on flat outcrop surfaces. Because you measure a greater strike length with this method, it gives the average strike on a larger scale than the contact method. Dip is often difficult to measure in some pavement exposures, because there may be little dip exposed. The end-on method must then be used; sometimes you may even have to lie down to do it. Move back a few metres, hold your clinometer at arm's length in front of you and align it with the trace of foliation seen in the end of the exposure, ensuring that your sight line is horizontal and in the strike of the plane being measured. Figure 6.5 shows an excellent exposure suitable for end-on dip measurement, but it can be used on far poorer exposures of dip than that.

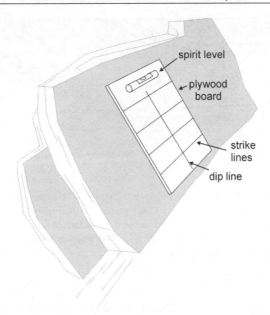

Figure 6.3 A simple aid for measuring strike and dip: an A4-sized plywood board permanently marked with a set of lines parallel to a fixed spirit level. These lines become strike lines once the board is levelled, and can be measured directly with the compass clinometer.

6.1.3 Method 3

This gives reliable measurements of strike and dip in regions where large areas of moderately dipping bedding planes are exposed or where surfaces are too uneven to measure in any other way. Extreme examples are the dip slopes often seen in semi-arid countries, but the method can also be used on smaller uneven surfaces, including joint planes. Stand at the end of the exposure (kneel or lie if necessary) and ensure that your eye is in the plane of the surface to be measured. Sight a horizontal (strike) line across the surface with a hand-level, then sight your compass along the same line and measure its bearing. This will give a reading that averages out the unevenness of the plane (Figure 6.6). To measure dip, move far enough back so that you can see as much dip surface as possible, then take an end-on reading (Figure 6.7). Compasses with built-in hand levels, such as the Brunton, are ideal to establish the strike line for this type of measurement.

Figure 6.4 *Measuring the strike of a veinlet on a rough horizontal surface by Method 2 (see text).*

Figure 6.5 *An ideal exposure for the end-on measurement of dip.*

Figure 6.6 *Measuring the strike of an uneven surface with a prismatic compass (Method 3; see text).*

Figure 6.7 *Measuring the dip of an uneven surface by Method 3 (see text).*

6.2 Plotting Strike and Dip

Plot dip and strike at the exposure as soon as you have measured them. This is very simple with a mirror compass, because as soon as the strike is measured it can be plotted as a symbol on the map in the same way as plotting a bearing (see Section 3.5.3). Briefly, take your strike reading and then, without disturbing the setting of the azimuth ring, align the orienting lines with a N grid line on the map and slide it into position.

If you have measured strike with a prismatic compass, plot it using the pencil-on-point method (Section 3.5.3).

6.3 Recording Strike and Dip

Whether you enter your strike and dip readings in your notebook as well as on your map is debatable, but if you lose your field map, you will have to start all over again from scratch anyway. It takes a little extra time, however, to record the strike and dip on the map against the strike/dip symbol. This is particularly convenient when mapping on aerial photographs when you must later re-plot your field information onto a base map on a different scale.

6.3.1 Conventions for writing down strike and dip

Like your field map, your notebook should be intelligible to others who may need to consult your work. Therefore readings should be written in a way that is not ambiguous. For example, strikes and dips must be recorded in a manner where there can be no possible confusion over the direction in which the planar structure dips. The most unambiguous way to do this is to write down three items in this format: 032/43 SE. The first item, 032, is the direction of strike (Figure 6.8a). Either one of the two possibilities, 032 or 212, can be given. Note that this is written deliberately with three digits, rather than just 32, to remind the reader of your notebook that this is a compass direction not an angle of dip. The second item, 43 in the present example, is the angle of dip and this is followed by the general dip direction expressed as a compass quadrant, here SE.

6.3.2 Alternative conventions

There are alternative conventions in use by geologists around the world that express the orientation of a plane by means of only two numbers;

1. *American right-hand-rule* (Figure 6.8b): If one views a dipping plane along strike it will appear to dip to the right or to the left depending on which of the strike directions you are using as the direction of viewing. Choose the direction of viewing that gives a dip to the right. The example above would be written as 032/43.

2. *British right-hand-rule* (Figure 6.8c): Place the right hand, palm down, on the dipping bed, with the thumb pointing down the dip. Record the strike in the direction of the index finger. The example above would read as 212/43.
3. *The dip direction/dip convention* (Figure 6.8d). Uses the direction of dip of the plane expressed as a compass direction. This gives 122/43.

Given the fact that these different methods exist, it is important that you state in the front of your notebook the convention used by you.

6.4 Measuring Linear Features

Linear features related to tectonic structures are termed lineations and the methods of measuring them described here can also be used for any other linear features, whether resulting from glaciation, currents associated with sedimentation or flowage in igneous intrusions.

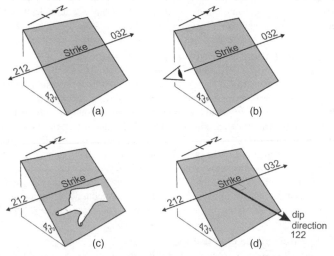

Figure 6.8 *Conventions for recording strike and dip. (a) Strike/dip/quadrant of dip direction, that is, 032/43 SE (or 212/43 SE). (b) Strike/dip; the strike direction is chosen that, when used as a viewing direction, gives a dip to the right, that is, 032/43. (c) Strike/dip; the strike corresponds to the direction in which the index finger of the right hand points when the thumb points down dip, that is, 212/43. (d) Dip direction/dip, that is, 122/43.*

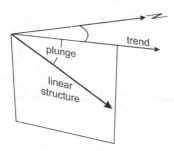

Figure 6.9 *The definition of trend and plunge of a linear structure.*

6.4.1 Trend, plunge and pitch (or rake)

A lineation is defined in space by its *trend* (the bearing of an imaginary vertical plane passing through it) and by its inclination, or *plunge*, in that plane (Figure 6.9). Some lineations appear as lines on an inclined geological surface, for instance where the trace of bedding can be seen on a cleavage plane.

Such lineations can often be measured more easily by their pitch (rake), that is, the angle the lineation makes with the strike of the surface on which it occurs (Figure 6.10a). Pitch can be measured with an ordinary transparent protractor, the bigger the better, or with a mirror compass using the method explained in Figure 2.11. This measurement has to be accompanied by the strike and dip readings of the plane on which the lineation lies. Log the pitch in your notebook

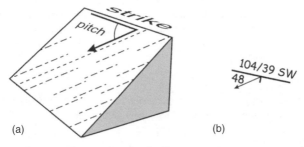

Figure 6.10 *(a) Geometry of pitch of a linear structure present on a dipping surface, for example, a foliation plane. (b) How to record pitch in your notebook by a diagram; record its angle together with the strike and dip of the surface that the linear structure lies on.*

Figure 6.11 *Stretched conglomerate pebbles in SW Uganda; trend and plunge can be measured directly.*

in this format: bedding 104/39 SW, pitch of lineation 48W. In this example the pitch is 48° and this angle was measured from the westerly end of the strike line (Figure 6.10). Provided the strike and dip of the surface have been measured, trend and plunge can be calculated using a stereographic net.

6.4.2 Measuring lineations

Although some lineations on a surface can be measured by their pitch, many must be measured directly as plunge and plunge direction, such as the lineation defined by the long axes of deformed pebbles (Figure 6.11). Figure 6.12 illustrates the measurement of plunge and trend. Firstly, the plunge is measured using the clinometer:

1. Rotate the azimuth ring to set up the clinometer's scale.
2. Open up the compass and align the long edge of the compass with the lineation ensuring that the compass is held vertical (Figure 6.12a).
3. Read the plunge on the clinometer scale.

Next the trend (plunge direction) is measured (Figure 6.12b):

1. Align the edge of the lid of the compass with the lineation.
2. Rotate the body of the compass about the hinge of the lid until the compass itself is level and so that the compass needle is free to rotate.

3. Holding the compass steady, rotate the azimuth ring until the orienting arrow lines up with the magnetic needle, as if taking a bearing.
4. The trend is read off against the bearing mark on the down-plunge side (labelled 'm' in Figure 6.12b). (Note that, unlike strike readings, there are not two possible readings 180° apart.)

Sometimes the trend can be measured by looking vertically down on the exposure and aligning the compass with the direction of the lineation in plan view. Lineations can be measured both accurately and easily by the Japanese compass illustrated in Figure 2.7.

6.5 Folds

Folds typically occur on a range of scales. Those with a large wavelength often display smaller, parasitic folds on their limbs (Figure 6.13). In spite of this difference in size, these two sets of folds are geometrically similar. They share the same orientation of their hinge lines and of their axial planes. This fact is useful for the mapping of folds that are too large to be seen as single exposures. The parasitic folds provide information on the orientation, and even the location, of major folds. We therefore explain now the essential measurements of folds.

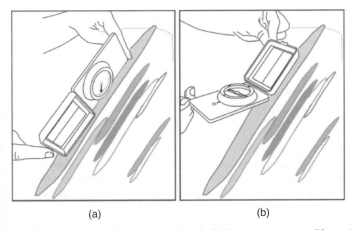

(a)   (b)

Figure 6.12 *Measuring the plunge and trend of a linear structure. (a) Plunge is measured with the clinometer held vertically and (b) trend is measured with the compass held horizontally.*

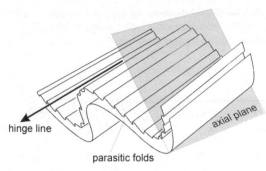

Figure 6.13 *Parasitic folds and larger-scale folds share similar orientations of their hinge lines and axial planes. The parasitic folds have different shapes on opposite limbs of the larger fold.*

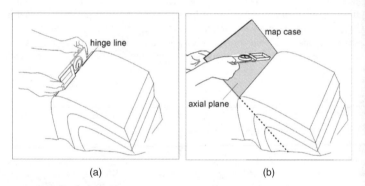

Figure 6.14 *Measuring minor folds. (a) The plunge of the hinge line and (b) the strike of the axial plane.*

6.5.1 Fold hinge lines

Measure the plunge and trend of the line on the folded surface that joins points of greatest curvature (Figure 6.14a). Use the method explained in Figure 6.12. Plotting these data on a stereogram will help with analysis (Lisle and Leyshon, 2004).

6.5.2 Axial planes

The axial plane is the surface geometrically containing the hinge lines of successive folded surfaces in a fold (Figure 6.14b). Such surfaces do not define planes that can be measured with the compass/clinometer. Therefore use your flat map case to provide a measurable plane (Figure 6.14b). Measure the strike and dip (Section 6.1). Such data will be needed to draw large-scale folds in your cross-section.

6.5.3 Fold asymmetry

Parasitic folds have different shapes at different positions within a major fold (Figure 6.13). S-shaped folds occur on one limb and z shapes on the other. This has obvious use in mapping large fold structures. For example, the z-fold seen in Figure 6.15 indicates that the major antiformal closure is to the right of the exposure, the synformal closure to the left. Record the fold asymmetry in your notebook, remembering to state the viewing direction (note that an s-fold becomes a z-fold if seen from the opposite direction). It also indicates inclination of the axial plane. Minor folds such as this are too small to show in an outcrop on your geological map except as a symbol selected from the list of symbols printed on the inside back cover of this book.

6.5.4 Folds with axial plane cleavage

Cleavages and other foliations formed during the folding of rocks usually adopt a special orientation in relation to the associated folds. Cleavage has an orientation approximately parallel to the axial plane of the fold (Figure 6.16), and this provides a useful tool for mapping fold structures; it uses the relative orientations

Figure 6.15 *Asymmetrical minor folds in Precambrian banded gneisses, Gjeroy, Norway. These z-folds indicate a major antiform to the right of the outcrop.*

Figure 6.16 Fold with axial plane cleavage in Devonian slates at Combe Martin, North Devon. North is to the left of the photograph.

of bedding and cleavage. At outcrops where both bedding and cleavage can be seen, make a field notebook sketch of their dips (Figure 6.17). Hold the notebook up to the outcrop and rotate it until cleavage appears to be vertical. When this is done, the dip direction of the bedding in the rotated drawing indicates the direction in which the nearest synform is to be found. For example, if the rotated bedding dips to the east, the nearest synform is located to the east of the exposure. If such observations are made repeatedly across an area and recorded on the map, large-scale antiforms and synforms can be mapped out.

The line of intersection of bedding and cleavage should be parallel to the hinge line of the fold. It is therefore a useful measure of fold orientation. Find the plunge and trend of this line by direct measurement (Figure 6.12) or construct it on a stereogram from the strike and dip readings of bedding and cleavage. For this construction see Lisle and Leyshon (2004). To record this on the field map, see the list of symbols at the back of this book.

6.5.5 Fold shape

One of the main aims of geological mapping is to discover the large-scale structure and to represent this by means of cross-sections, and so on. Wavy map patterns do not necessarily indicate the presence of folds; unfolded rocks can appear folded on the map because of topographic effects. Changes of bedding

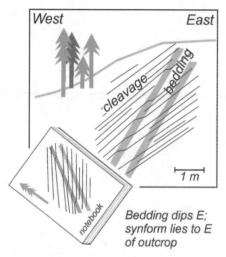

Figure 6.17 *Mapping out fold structures from cleavage and bedding. Make a notebook sketch of the dipping bedding and cleavage, hold it up to the outcrop and rotate until the cleavage appears to be vertical; the dip direction of the bedding now indicates the direction of the nearest synform.*

dip, of younging direction, of asymmetry of minor folds or of bedding/cleavage relationships may indicate the existence and position of large-scale folds in your mapping area. But what style will these large-scale folds have? The most reliable indication for this is the shape of small folds seen at the outcrop. Look closely at the curvature of the inner and outer arcs of the folded layer (Figure 6.18). Is the curvature of the inner arc greater, equal to or lower than the curvature of the outer arc (corresponding to fold classes 1, 2 and 3, respectively)? If the fold belongs to class 1, is the layer thickness in the hinge greater, equal to or less than the thickness on the fold limbs (corresponding to classes 1A, 1B and 1C, respectively)? Look for a change of fold shape with lithology. When you draw the cross-section, draw folds with the type of shapes seen on a small scale in the field.

There is an extensive terminology for the description of folds; therefore before going into the field you are well advised to read Fleuty's (1964) paper or to take McClay's (2003) handbook on your fieldwork. Show with symbols the trends, plunges and shapes of all folds too small to show in any other way. Make notebook sketches.

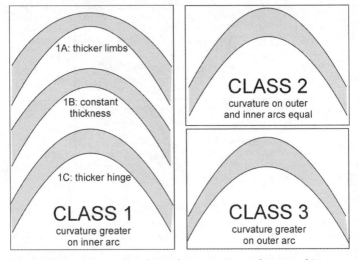

Figure 6.18 Fold classes based on relative curvature of outer and inner arcs and layer thickness.

Data checklist for folds

- Type (antiform, synform, recumbent).
- Fold hinge line orientation (plunge, plunge direction; see Figure 6.14a).
- Axial plane (e.g. strike, dip; see Figure 6.14b).
- Symmetry when viewed in the plunge direction (s, z, m; see Figure 6.13).
- Direction to larger-scale antiform indicated by asymmetry or bedding cleavage (Figure 6.17).
- Fold class (Figure 6.18).
- Associated structures (cleavage, jointing, veining).
- Sketch/photograph taken in the down-plunge direction.

6.6 Faults

Most small faults never get mapped because they are never seen. Many have small displacements that it matters little if they are individually missed, but record those you do see in your notebook to help you to establish a fracture pattern. Major faults are more likely to be found, but even those with displacements of tens of metres may be missed where exposure is poor. Compare the multiplicity of faults on a coalfield geological map with those on a map from non-coal-bearing area. The ground is probably just as faulted on both maps. But in the coalfield the faults have been detected underground and projected to the surface. Many faults have to be mapped by inference. Suspect a fault where:

- there are unaccountable changes in lithology;
- sequences are repeated;
- part of the sequence is absent;
- strikes of specific beds cannot be projected to the next exposure;
- joint spacing suddenly decreasing to a few centimetres; or
- a zone of veining occurs.

Topography is often a good guide. Faults may result in spring lines, boggy hollows, seepages or, in semi-arid countries, a line of taller greener trees, flanked by lower flat-topped acacia. However, beware; although most fault zones erode a little faster than the adjacent rocks to form longitudinal depressions, some faults in limestones may form low ridges owing to slight silicification, which helps to resist erosion. Faults are more easily traced on aerial photographs, where the vertical exaggeration of topography seen under a stereoscope accentuates those minor linear features called lineaments, features often to find on the ground; many of them are probably faults.

The *slip* (real displacement) on faults is usually difficult to deduce. What can often be seen is a *separation* – an offset of beds on maps or sections. Separation is not related to slip in any direct way. For instance, a fault on a map with dextral strike separation – that is, where the beds on the opposite block are shifted to the right relative to those in the block you are standing on – does not imply that the slip has a strike component that is dextral (Figure 6.19a). Therefore avoid the use of arrow symbols along the fault; they give the false

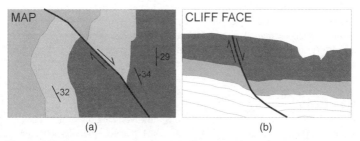

(a)   (b)

Figure 6.19 Separation: (a) map of fault with strike separation of dextral sense and (b) cliff section showing faulted units. In both cases the arrows are confusing because they suggest the sense of slip (true movement), which here is not known.

impression that the slip of the fault is known. Likewise, a fault seen on a steep outcrop surface and showing dip separation may be the result of pure strike-slip (Figure 6.19b). The recommendation therefore is to use the term separation, unless enough information is found to deduce the net slip.

In textbooks much is made of slickenlines, and if they are seen they should be noted, but do not put much faith in them, they merely reflect the last phase of movement. Most faults have moved several times, although not always in the same direction. Note also that faults may have a thickness wide enough to show on the map. Faults may also be breccia- or gouge-filled, or even mineralised, perhaps with calcite or even fluorite. Note such facts in your notebook.

6.7 Thrusts

Thrusts are low-angle reverse faults. They can be very large and important structures, but sometimes can completely escape notice. They often become more obvious from the map pattern (Figure 6.20). If the local stratigraphic ages are known, large thrusts are usually obvious since they result in older rocks overlying younger ones. Some thrusts are made up of flats and ramps; the former are thrust segments that track along parallel to formation boundaries whereas the latter are parts of the thrust that cut obliquely through the contacts (Figure 6.20). It is important to note that these relationships are visible on the map and well as in the cross-section.

If the thrust surface itself is exposed, the position should be clearer. There may be shearing along the surface, or there may be *mylonite*. Where mylonite does occur it may be thick enough to map as a formation in itself and form a useful marker. The lower part of the upper plate should not show any of the sedimentary features you would expect in a stratigraphic unconformity. However, not all thrusts are major thrusts. They may occur in imbricate zones, consisting of numerous small sub-parallel thrusts associated with major thrusts, as in the Scottish Moine Thrust Zone. Such zones are marked by multiple repetitions of partial sequences that, if poorly exposed, are impossible to map completely. Sometimes the spacing between individual thrusts may be only a few metres, sometimes tens of metres.

6.8 Joints

Joints, like faults, are rock fractures. Joints, however, lack discernible displacement. They occur in every type of rock – sedimentary, pyroclastic, plutonic, hypabyssal, volcanic and metamorphic. Do record joints, but do not clutter your map with them. Enter them into your notebook and later plot them on transparent overlays to your fair copy map, or plot them as statistical diagrams, such as stereograms and rose diagrams in equal-area 'cells' spread over the surface of

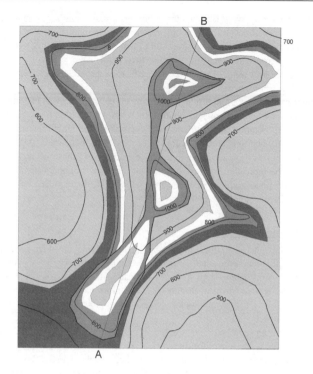

Figure 6.20 *Mapping a thrust. Over much of its outcrop, the thrust forms flats where there is concordance with the formation boundaries making it difficult to recognise as a thrust. The presence of a ramp where the thrust plane cuts through the units at an angle and the repetition of units helps with its recognition.*

your map overlay. Master joints, those dominant major joints, are an exception. They can sometimes warrant being shown on your map. Follow them on the ground or on aerial photographs, and plot them in a similar manner to faults, but with the appropriate 'joint-dip' symbol. In general, keep joints off your maps, but do not forget them. These are important to water supply, pollution control and hydrocarbon reservoirs.

Measure the strike and dip of joints in much the same way as bedding. Often their surfaces are uneven and contact methods unsuitable. Book readings in your chosen notation and estimate, where possible, the length and the spacing of joints in each set, and what formations each set penetrates. Master joints may show up well on aerial photographs, especially in limestone regions where they may be indicated by karst patterns and lines of sinkholes (*dolines*). Joint patterns on photographs can sometimes be used to distinguish one formation from another.

6.9 Unconformities

Stratigraphic unconformities show younger rocks lying on older rocks below, but their junction represents a break in sedimentation. The rocks just above an unconformity often show features indicating that they were deposited on an already eroded surface (Figure 6.21). Unfortunately, this relationship is not always as clear as textbooks suggest, especially where rocks have been metamorphosed.

The break in sedimentation is easier to demonstrate when the older rocks are tilted relative to the unconformity surface producing an *angular unconformity*. The contrasting dips and the truncation of the older beds make this obvious at the outcrop (Figure 6.22). Similar angular relationships should be present on the map scale too. Sometimes, to confuse matters, there is angular unconformity on both sides of the break if the later rocks were deposited on a sloping surface. A *disconformity* may be even more difficult to recognise; the beds are parallel both above and below it. It should be discovered during sedimentary logging by evidence of erosion between the two stages of deposition.

6.10 Map Symbols

There are internationally accepted geological map symbols; unfortunately every national organisation has its own interpretation of them. A short list of the main symbols is printed on the inside back cover of this book. Berkman (2001) gives a comprehensive list spread over many pages. However, there may be situations where an improvised symbol has to be resorted to; just make sure that the map key explains it.

Note, a strike symbol is drawn on a map with its *centre* at the point where the reading was taken. The point of a lineation arrow head is the point where

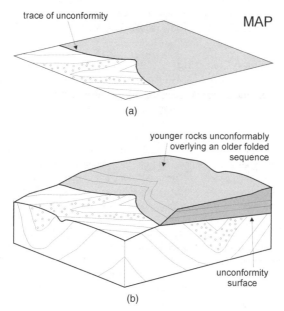

Figure 6.21 An angular unconformity displayed (a) on a geological map and (b) in a block diagram.

Figure 6.22 An unconformity at Sully Island, South Wales. Triassic mudstones rest unconformably on dipping Carboniferous limestone.

that reading was made. The exception is where several readings are made at one point: in that case the symbols radiate from the observation point.

6.11 Specimen Collecting

Collect representative specimens of every formation and rock-type you show on your map. Often, several specimens of the same formation are required to sample the variation in composition. Some variations of composition may not, of course, be obvious in a hand specimen so extra specimens are needed as a safeguard. The size of specimen you collect must depend on the purpose you wish to put it to, not on what you think you can carry. Before you go into the field, consult your rock cutter to find out the adequate size for thin-sectioning. Whenever possible, choose specimens showing both weathered and unweathered surfaces, and if necessary collect two specimens to show both aspects. Do not collect just a piece of rock you can knock off an exposure with a hammer. The easiest piece of rock to break off may not be representative of the exposure as a whole. You may have to spend considerable time in breaking out a good specimen with hammer and chisel. Having broken off a specimen, trim it. Mark sedimentary rock specimens to show which is their top.

Metamorphic specimens may need to be oriented so that directional thin-sections can be cut. A simple way of recording the orientation of collected hand specimens is suggested by Prior *et al.* (1987):

1. Choose a flat surface on the rock to be sampled (Figure 6.23).
2. Measure the surface's strike, dip and dip direction (N, S, E or W) and record this in your notebook (Section 6.3).

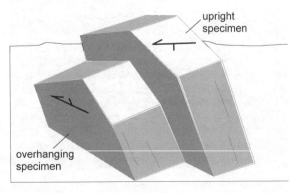

Figure 6.23 Marking a specimen to record its in situ orientation.

3. Draw a strike line on the chosen surface with a half-arrow to indicate the direction of the recorded strike, with the barb pointing in the upward direction.
4. Draw a short tick on one side of strike line; on the lower side if the selected surface faces upwards, and on the upper side if the surface faces downwards, that is, if the surface is overhanging.
5. The specimen can now be removed.

6.11.1 Marking specimens

Marking rough, wet and often friable rock specimens with a unique and permanent specimen number is often not a trouble-free task. Light-coloured, smooth specimens are best marked with a black waterproof permanent marker pen. Dark-coloured rocks can be numbered in the field by painting on a unique number with permanent quick-drying 'Tippex' (make sure it's not the water-soluble kind). Yellow electrical tape with a number written on it can be used as a temporary solution until you get back to base camp. Remember to wrap each specimen in a plastic specimen bag or newspaper to protect it from bruising in your rucksack, and incidentally to protect your field equipment from scratching and dirt. In camp, wash your specimens to clean them and to remove any loose surface material, then dry them. When dry, paint on a small patch using white or yellow enamel model paint, and when that's dry, number the specimen by using a fine permanent black marker pen. If you do not want to carry paint pots and small brushes into the field, then a much easier solution is to use liquid paint marker pens (e.g. Tamiya or Edding) available from specialist websites or model shops. Once all the paint is dry, wrap the specimens in the specimen bag or newspaper and number each packet on the outside with a felt-tipped marker, so that you can easily locate any specimen you might wish to look at.

6.11.2 Samples of fossils

Some fossils are easy to remove from their rock matrix, others are not. Many are deeply embedded with only a small portion showing; scrape away enough rock with a knife to see whether the specimen is worth collecting, and if so then break out the rock containing it. Many fossils are casts or impressions in the rock; again collect the piece of rock containing them. Wherever possible, collect both external and internal casts: both are important. Sometimes you may need to collect several kilograms of fossiliferous rock so that individual fossils can be extracted in the laboratory. This is particularly so where microfossils are needed. Mark all specimens with the way-up in which they were found.

Pack delicate specimens in boxes or tins and pad them with cotton wool, tissue paper or newspaper or use expanded polystyrene ceiling tiles cut to fit the boxes. Use grass if there is nothing else. Carry a selection of boxes, from matchbox upwards in size. Wrap non-fragile specimens in newspaper and treat them in the same way as rock specimens.

As with rock specimens, do not collect more than you need; do not clean out a good locality to sell to dealers, and report anybody you see doing so – this applies equally for mineral localities.

6.11.3 Booking specimens

Log specimens in your notebook immediately after you have collected them. Preferably, write the specimen numbers in the left-hand margin of the page so that their details can be found easily. If their numbers are written with red pencil, they can be even more easily distinguished from field observation numbers in the same column. Alternatively, if you are collecting large numbers of specimens, add a column to your notebook specifically for specimen numbers. In addition to logging specimens on the working pages of your notebook, register them with a brief description in a specific index at the back of your notebook too. This avoids finding yourself with two almost identical specimens from different places with the same number, and no way to tell which is which. A register also helps you to ensure that you have collected specimens of everything you should have collected, and if you give the notebook page numbers where they are more fully described, it acts as a handy ready-reference (Figure 6.24).

6.11.4 Transporting specimens back to the lab

Geological specimens are heavy and if shipped in a box that is too large can only be accepted as freight. Smaller boxes, which one can lift easily, can go much more quickly by passenger transport. A box about $25 \times 30 \times 25$ cm made of timber about 1 cm thick, battened and steel banded, is acceptable by TIR rules, railway passenger services and airlines. Mark your name and address on top and on at least one side, and add: ROCK SPECIMENS FOR SCIENTIFIC RESEARCH.

Never write 'ore specimens' or 'mineral specimens' on boxes or on customs declarations. Most countries do not appear to have export regulations controlling 'rocks' but do so for *minerals* and *ores*. The honest declaration of geological materials as 'rocks' avoids bureaucratic delays, and gets your samples back in your laboratory more quickly.

6.12 Field Photography

A camera is an essential tool for a field geologist. Since the last edition of this text in 2004, digital photography has totally replaced film-based photography for field geology. Therefore we will assume that the geologist will be carrying a digital camera. You will need to capture many images to remind yourself of field landscapes, rock outcrops or close-ups of small specimens. Many images will be

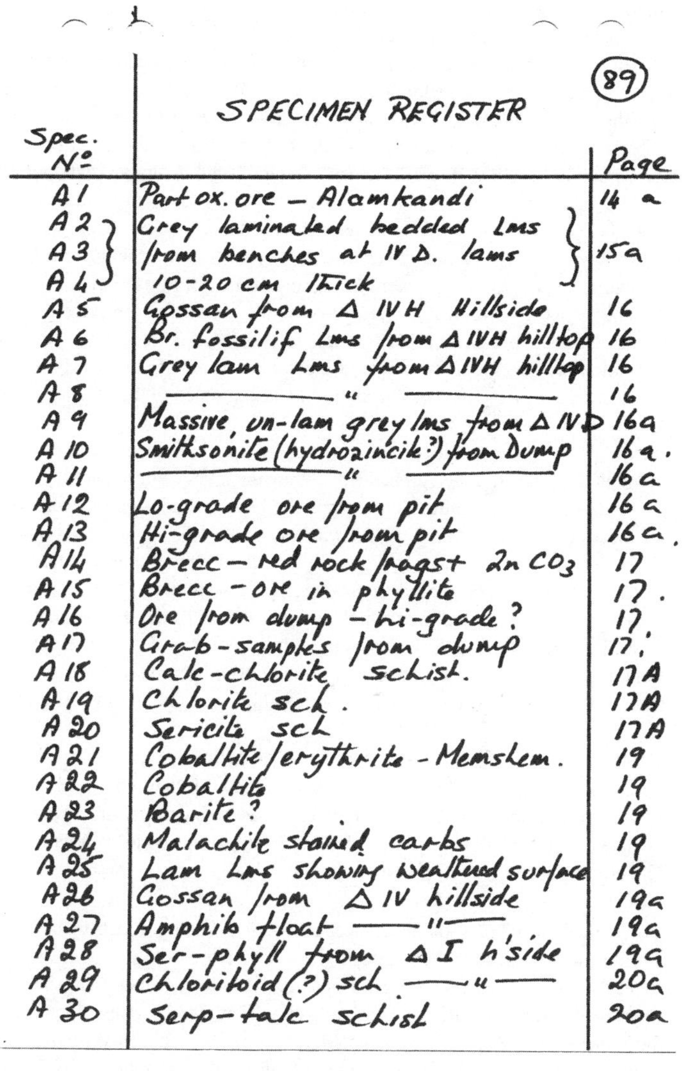

(89)

SPECIMEN REGISTER

Spec. №		Page
A 1	Part ox. ore – Alamkandi	14 a
A 2	Grey laminated bedded Lms	
A 3	from benches at IV D. lams	15a
A 4	10-20 cm thick	
A 5	Gossan from △ IVH Hillside	16
A 6	Br. fossilif. Lms from △IVH hilltop	16
A 7	Grey lam Lms from △IVH hilltop	16
A 8	"	16
A 9	Massive, un-lam grey lms from △IVD	16a
A 10	Smithsonite (hydrozincite?) from Dump	16 a.
A 11	"	16 a
A 12	Lo-grade ore from pit	16 a
A 13	Hi-grade ore from pit	16 a.
A 14	Brecc – red rock fragst + Zn CO₃	17
A 15	Brecc – ore in phyllite	17.
A 16	Ore from dump – hi-grade?	17.
A 17	Grab-samples from dump	17.
A 18	Calc-chlorite schist.	17A
A 19	Chlorite sch.	17A
A 20	Sericite sch	17A
A 21	Cobaltite/erythrite – Memshem.	19
A 22	Cobaltite	19
A 23	Barite?	19
A 24	Malachite stained carbs	19
A 25	Lam Lms showing weathered surface	19
A 26	Gossan from △ IV hillside	19a
A 27	Amphib float ——"——	19a
A 28	Ser-phyll from △ I h'side	19a
A 29	Chloritoid(?) sch ——"——	20a
A 30	Serp-talc schist	20a

Figure 6.24 A specimen register in a field notebook.

required to illustrate your technical report and perhaps an interesting Powerpoint talk about your fieldwork programme. It is very easy with digital photography to take thousands of field photographs and rely on your own memory regarding where each was taken; this technique is usually disastrous. Treat photography as part of the data collection process and write down in your notebook the frame number, geographical location and view direction of at least the key important images. Back at base it is often difficult to recognise the geological features for which the shot was taken. Therefore whenever you take an important image, draw a labelled sketch of the same scene in your notebook. When photographing rock exposures, always include a scale; for a large exposure, include a human; for a smaller exposure or specimen, use a ruler/scale. Only use coins as a last resort; they vary in size from country to country. You can usefully carry a blue plastic sheet to use as a background for photographing hand specimens or borehole cores. Remember to record the specimen details in each image; it is best to use a whiteboard pen on some blank laminated white A5 size card, which can be wiped clean to be used again.

Make sure your camera clock is set correctly so that the date and time of every photograph will be digitally embedded within the image file. Some cameras have a plug-in GPS capability that records the latitude and longitude of every photo location, although this is in no way essential. The key to good-quality photographic images is always sunlight; the best field photographs are taken on sunny days with the sun behind the photographer's shoulder. If you want to photograph an important cliff section, then the best time of day for photography is either the early morning or late evening when the sun is low in the sky and illuminating the cliff face (Figure 6.25). Photos taken at midday, when the sun is vertically overhead, usually result in poorer, high-contrast fieldwork images.

6.12.1 Using digital cameras for field photography

If you are thinking of purchasing a camera for fieldwork then you need to consider what types of photographs you are likely to take, since there are advantages and disadvantages of digital systems that the geologist needs to be aware of.

The advantages of digital photography are:

- An immediate display of the image, allowing a quality check to be made before moving on to the next locality.
- Because the image is stored electronically, the digital image can be later improved using software such as Photoshop Elements in terms of cropping, brightness, contrast, sharpening, colour hue, and so on. Once modified to bring out better the features of interest, the image can be easily pasted directly into the computer file containing your report or dissertation.

Figure 6.25 *Example of a good field photograph; folded upper Jurassic-lower Cretaceous rocks, Stair Hole, Dorset. A 70 mm focal-length lens is used to avoid wide-angle distortion. The sun is shining directly on to the cliff face. People and a fence give the scale.*

The disadvantages of digital cameras for fieldwork are:

- The field ruggedness of many digital cameras remains a consideration; like all electronic instruments they do not like moisture, and sea water is fatal. It is not recommended to buy an expensive system with interchangeable lenses for active fieldwork because dust and moisture may enter the camera's electronics and totally disable the system. If possible, use a UV filter on the front of the lens to protect it from dust and damage.
- It is yet another electronic device that you are carrying into the field that will require battery power, either rechargeable or disposable.
- Images are stored on memory cards that have finite storage capacity. You will either need a bank of cards safely stored back at base camp or a secure field laptop computer onto which you can download each day's set of images.

There is a bewildering array of ever-changing digital camera systems on the market ranging from £40 to over £1000, and most mobile phones also have an internal camera. However, the cheaper end of the range and mobile phones are targeted at the 'point and shoot' market with total automation, which is not ideal for field photography. The size and quality of the digital image is one consideration. You only need a 2 megapixel image to produce a full-screen Powerpoint display, but a good field camera should be at least 6 megapixel to allow for some image editing flexibility. Most cameras store images as compressed JPEG images, whereas better quality cameras also have the capability of storing images as uncompressed TIFF or RAW formats. Although these uncompressed files are large, they do allow for higher quality technical images to be produced by digital post-processing.

A field camera should have a zoom capability so that you can use the wide angle to photograph large cliff sections, a 1:1 image ratio setting to allow you to take accurate images of outcrops without any lens distortion, plus a telephoto setting for photographing inaccessible cliff sections. All automatic cameras make the assumption that the average tone of any image is mid-grey and set the exposure accordingly. In field geology you may be photographing basalt (black) or chalk (white) and this mid-grey assumption is just not valid. For this reason a good field camera should have the ability to override the automatic exposure settings. At high altitudes or in snowy conditions there is often a lot of haze in the atmosphere caused by high UV levels and light scattering. Using a UV or polarising filter on the front of the camera lens can dramatically improve landscape photography in hazy or snowy conditions.

Whilst in the field, you will also need to photograph small specimens, such as fossils, perhaps only a centimetre in size. When photographing any close-up specimen remember to include a millimetre ruler scale in the image. The capability of any digital camera to take quality macro (close-up) photographs is highly variable. When testing cameras for field suitability, take a close-up photograph of a £1 coin and look carefully at the image quality produced.

One very useful exercise in field geology is to photograph a rock outcrop for subsequent sketching of a cross-section. If you set your camera to wide-angle in order to fit the whole outcrop in a single picture, the result will be a low-quality image with severe distortion effects. A much better technique is to produce a 'photo-mosaic' that also includes a scale (Figures 6.26 and 6.27). Firstly carry a small selection of drawing chalk of different colours (white, yellow, red) and with a tape measure clearly mark the exposure with chalk marks of contrasting colour at regular intervals of 1–10 m depending on the outcrop size. Set your camera zoom lens to 1:1 magnification; this is equivalent to a 70 mm focal length lens for 35 mm film cameras. Stand back from the outcrop and use the camera in horizontal (landscape) or vertical (portrait) format depending on the outcrop height. Take a series of images by walking parallel with the outcrop, making absolutely sure that each overlaps with the previous image by at least 25%. Try to include two chalk marks in each image. A similar process can be carried out from a boat for sea cliff sections by sailing parallel to the cliff

Figure 6.26 *Single digital image of thrusted sediments exposed at Dinas Dinlle, North Wales coast. Note the use of a 3 m survey staff as a scale.*

Figure 6.27 *Dinas Dinlle photo mosaic constructed from six overlapping digital images with 10 m markers overlain. This mosaic can be used to construct a technically accurate sketch of the outcrop.*

and taking continuous overlapping images. For a scale, a tape measure can be subsequently used to measure the width of houses or the distance between two objects visible on the sea cliff photo-mosaic. Using one of the many available photo-mosaic software packages you can subsequently stitch or mosaic all these images together to produce a single long thin image that has minimal distortion. The best stitching software makes a distortional lens correction before stitching the images. Using the chalk marks or the distance between two identifiable points, you can construct a scale so you can use this 'photo-mosaic' image as a basis for a technically accurate section.

123

6.13 Panning

To be able to use a gold pan is a useful skill for the geologist, but it needs practice. Gold and cassiterite can be panned from streams, but many dense rock minerals that survive erosion can also be concentrated by panning. These include garnet, rutile, zircon, epidote, monazite, magnetite, haematite and ilmenite. Differences in the 'heavy mineral suites' extracted by panning soils are useful guides to the underlying geology in poorly exposed regions (see *loaming*, Section 4.6.3).

Because of its density, native gold (specific gravity [SG] 16.6–19.3 depending on its purity) is easy to concentrate in a pan, but garnet and epidote (SG 3.2–4.3) are only a little denser than sand and rock debris (SG 2.7) and more skill is needed to concentrate them. A 30 cm diameter metal or plastic pan is sufficient for purely geological purposes. Keep it spotless and free from rust and grease, so do not use it as the camp frying pan. Collect *stream gravel* from the coarsest material you can find, for that is where the heavier minerals concentrate (Figure 6.28). Dig for it with a trowel or entrenching tool and get down to bedrock if possible. Collect *soils* from below the humus. Heap the pan full of material, and then shake it vigorously underwater in a stream, or even in a tin bath. The finer heavies will pass down through the coarser light material, a process known to the mineral dresser as *jigging* (and this is what, with the

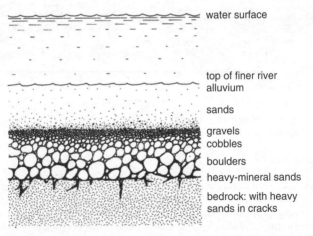

water surface

top of finer river alluvium

sands

gravels
cobbles

boulders

heavy-mineral sands

bedrock: with heavy sands in cracks

Figure 6.28 *A simplified profile through stream gravels; the heavy-mineral sands accumulate at the base of the coarser material and may even penetrate cracks in the bedrock. Note the sands and gravels of the stream are in constant movement, which allows the 'heavies' to pass downwards through them.*

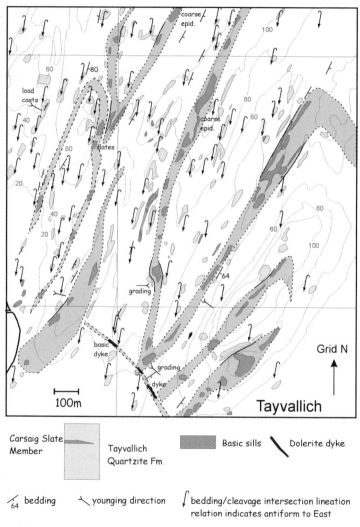

Plate 2 *An example of a field slip, Tayvallich area, Argyll, Scotland.*

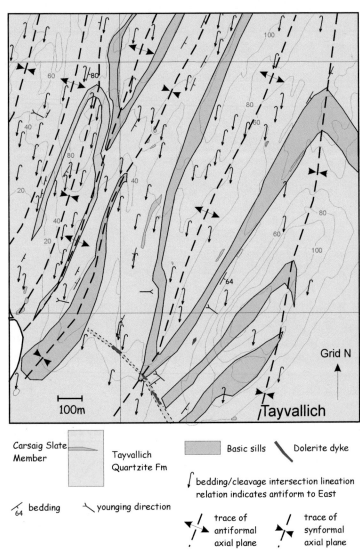

Plate 3 *The fair copy map of the Tayvallich area for comparison with the field map in Plate 2.*

Figure 6.29 A geologist panning in the Euphrates Valley.

constant movement of streams, helps to concentrate the 'heavies' on bedrock). Larger pebbles can be scraped off the top and discarded. Gradually wash off the finer, lighter material by: tilting the pan, dipping it into water; lifting it out, swirling it around; and dipping again; until only a small amount of usually darker and sand-size material is left. Then, using only a small amount of water, give a final careful swirl with the pan at an angle of about 30° to form a *tail* of sediment, graded with the 'heavies' at one end in order of their densities, leaving the lighter material at the other end (Figure 6.29). Now, under a cover of a little water, identify any minerals you can, using your hand lens. Wash the concentrates into a phial for future examination, using a plastic funnel and a camel-hair brush. Label the phial. Decant the surplus water back in camp. Panning is like fishing: you do not have to find anything to enjoy it.

7

MAPPABLE ROCK UNITS AND LITHOLOGY

In Section 4.2 we stated that the basic rock unit for geological mapping is the formation. The concept of the formation is so important that we need to go into more detail here. The geologist has the task of dividing the rocks of the mapping area into formations, and must decide on the number of formations, their thickness and their boundaries. Until this is done, or at least partly done, little progress can be made with the geological mapping in a cartographic sense. Obviously, exposures cannot be coloured in on the map until formations have been decided upon and colours assigned to them. For this reason, the initial stages of mapping can be frustrating, with little to show for the work done in terms of map coverage. The job of defining formations therefore clearly has priority over others. Below we give you advice with this.

However, it is important to realise that mapping involves more than colouring in a map to show the distribution of outcrops of the various formations. Remember it is mainly for practical reasons that the huge variety of rocks that crop out in an area are assigned to a few formations. A list of formation names conveys very little about the geology of an area. Mapping must therefore also involve the recording of detailed lithological information from within the various formations. This chapter suggests what information has to be gathered and how it can be recorded.

It is assumed that readers are already familiar with systematic methods of rock naming in the laboratory, know basic palaeontology and can recognise the mineral materials mentioned. Here you are told how to apply that knowledge in the field.

7.1 Lithostratigraphy and Sedimentary Rocks

Earlier we used the word formation in a very general sense to mean a mappable rock unit. However, for formal use there is an accepted conventional lithostratigraphical hierarchy of terms to describe the grouping of rock units (Holland et al., 1978, p. 8):

- supergroup
- group

Basic Geological Mapping, Fifth Edition.
Richard J. Lisle, Peter J. Brabham and John W. Barnes.
© 2011 John Wiley & Sons, Ltd. Published 2011 by John Wiley & Sons, Ltd.

- formation
- member
- bed.

7.2 Sedimentary Formations

A sedimentary formation has internal homogeneity, or distinctive lithological features that constitute a form of unit in comparison with adjacent strata. It is the basic local mappable unit. It crops out and can be traced sub-surface to other exposures; you show it on your map with a distinctive colour. It is the primary local unit (Holland *et al.*, 1978). If a formation has not already been formally named, name it yourself in the approved manner, attaching a place name to the rock name, for example *Casterbridge Limestone Formation*, or some geographic name, for example *Entrada Sandstone Formation* named after Entrada Point in Utah. Avoid formation names such as *White Limestone* or *Brachiopod Bed*. Establish a type section for every named formation for reference. See the Geological Society of London's guide on the subject (Geological Society, 1972). The US Geological Survey offers similar advice (Cohee, 1962).

Sometimes one or more *members* may be defined within a formation. A member has some distinct lithological character but may not be continuous. The smallest division of a formation is a *bed*, which is a unit with a well-marked difference from the strata above and below it.

A group consists of two or more naturally related formations. A supergroup consists of two or more associated groups. Groups do not have to be collected into supergroups.

7.2.1 How to define your formations

In the early days of mapping use the traversing method (see Section 4.3) to examine exposures on sections running across the strike of the rocks. Choose well-exposed sections such as streams or road cuttings. Record the characteristics of the rocks, paying attention to potential subdivisions within the sequence. Since the essential feature of a formation is its 'mappability', consider the scale of map and minimum thickness of the formation that will allow it to be drawn. Roughly speaking, anything narrower than 2 mm on the map is difficult to plot. For a 1:10 000 scale map, a formation would need to be at least about 20 m thick. Sometimes, though, a thin but highly distinctive unit may warrant mapping as a separate member because of its value as a marker.

Another aspect of mappability is whether the chosen boundaries of the formation will be traceable across country. When considering defining a boundary between two formations in your traverse take a look at the less well-exposed ground nearby and assess whether the boundary will be easily followed across the area. With this in mind, choose contacts between rocks with contrasting resistance to weathering. Such contacts are likely to be expressed by the geomorphology, and indicated by topographic features (see Section 4.5.3).

When deciding on a formation boundary, establish precise criteria for identifying the contact. These criteria will be used during the subsequent mapping and need to be stated explicitly in your mapping report. An example is taken of the Burmeister Formation, Kaka District, New Zealand (Johnston, 1971):

> The base of the formation is defined as the base of the graded bedded sandstone-siltstone beds and the top is defined as the last appearance of a sandstone bed.

7.3 Rock Descriptions

The definition of formations is an essential stage in the mapping process. When you have completed this task, the mapping can progress by establishing the position of contacts on your map. Exposures still need to be visited and the rocks described. Descriptions of the range of rock-types that make up the various formations are essential when you come to write your report later. A rock description made from memory, perhaps weeks later, is unlikely to be accurate or complete. One made in the field describes the rock as seen, with measurements of specific features, and factual comments on those subtle characteristics that are impossible to remember properly later. It also ensures that you record *all* the details needed.

Systematically describe in turn each rock unit shown on your map. Preferably work from the general to the particular. Describe first the appearance of the ground it covers: its topography, vegetation, land use and any economic activity associated with it. If the soils are distinctive, describe them too. Next describe the rock exposures themselves: their size, frequency and shape; whether they are turtlebacks, pavements or tors; or rounded or jagged ridges, gentle scarps or cliffs. Comment on joint spacing, bedding laminations, structures, textures, cleavage and other foliations. Support your observations with *measurements*. Describe the colour of the rock on both weathered and freshly broken surfaces. Weathering often emphasises textures; for example, the honeycomb of quartz left on the surface of granites after feldspars have been leached away immediately distinguishes silicic from less silicic varieties. Finally, describe the features seen in a hand specimen, both with and without a hand lens. Note texture, grain size and relationship between grains. Identify the minerals and estimate their relative quantities (see Appendix D, Chart D.1), bearing in mind the tendency to overestimate the proportion of dark minerals over paler varieties. Name the rock. Where appropriate, prepare a sedimentary section and/or log (see Section 7.4.3). Remember, that you can take a specimen home with you, but not an exposure. Ensure that you do have all the information you need before you leave the field.

7.4 Identifying and Naming Rocks in the Field

There are two problems here. The first is to find out what the rock is in pet-rographic terms; the second is to give it an identifying name to use on your fair copy map and in your report. The first is the *field name*, the second the *formation name*.

7.4.1 Field names

A field name should be descriptive. It should say succinctly what the rock is, but you cannot name a rock until you have identified it. A field geologist should, using a hand lens, be able to determine the texture and the relationship between minerals, and estimate the relative abundances in most rocks. He or she should be able to distinguish plagioclase from orthoclase, and augite from hornblende in all but the finer-grained rocks. She or he should be able to give some sort of field name to any rock. Dietrich and Skinner's *Rocks and Rock Minerals* (1979) is an excellent guide to identifying rocks without a microscope. Other useful guides are those of Tucker (2003), Stow (2005), Bishop *et al.* (2003), Fry (1984) and Thorpe and Brown (1991).

A field name should indicate structure, texture, grain-size, colour, mineral content and the general classification the rock falls into, for example *thin-bedded fine-grained buff sandstone* or *porphyritic medium-grained red muscovite granite*. These are the full field names, but shortened versions, or even initials, can be used on your field map. Avoid at all costs calling your rocks A, B, C, D and so on, on the assumption that you can name them properly in the laboratory later: this is the coward's way out. If you are really stuck for a name, and with finer grained rocks it does happen, then call it spotted green rock, or even *red-spotted green rock* to distinguish it from *white-spotted green rock*, if need be. Ensure, however, that you have a type specimen of every rock named. Sometimes you may find it helps to carry small chips around with you in the field for comparison.

7.4.2 Stratigraphic sections

Stratigraphic sections show the sequence of rocks in a mapped region, distinguishing and naming the formations and members that comprise them. They show the thickness of the units, the relationships between them, any unconformities or breaks in succession, and the fossils found. It is impossible to find one continuous exposure that will exhibit the complete succession of a region (even in the Grand Canyon), and a complete succession is built up from a number of overlapping partial sections. There may even be gaps where formations are incompletely exposed.

Sections can be measured in a number of different ways, and some guidelines are given here. The first task is to find a suitable place with good exposure. Make measurements of the true thicknesses of the beds, starting at the base of

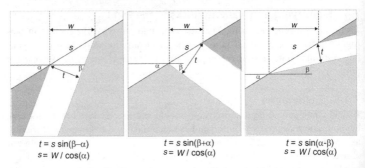

$t = s \sin(\beta-\alpha)$
$s = W / \cos(\alpha)$

$t = s \sin(\beta+\alpha)$
$s = W / \cos(\alpha)$

$t = s \sin(\alpha-\beta)$
$s = W / \cos(\alpha)$

Figure 7.1 *Calculation of stratigraphic thickness,* t, *from down-slope exposure distance,* s, *or width of outcrop,* W. α *and* β *are the angles of ground slope and formation dip, respectively. The values of* s *and* W *are measured at right angles to the strike.*

the sequence, and log them in your notebook as a vertical column. In measuring thickness, corrections must be made for the dip of the beds and the slope of the surface on which they crop out. This can be done graphically or trigonometrically (Figure 7.1 and Ragan, 2009). Compton (1985) illustrates several methods for measuring true thickness directly.

Indicate on the stratigraphic section the name and extent of every lithological unit, together with the rock-types within it. Take specimens of everything logged. Mark and note the names of any fossils found; collect specimens for later identification where necessary. Note the position of every section on your field map. Back in base camp, redraw to scale the sections from your notebook on squared paper. Later the section may be simplified and combined with sections from other parts of your mapping area as a columnar section or a fence diagram (see Sections 10.5 and 10.6.3). Stratigraphic sections may also include igneous and metamorphic rocks.

7.4.3 Sedimentary (graphic) logs

Although there are similarities, sedimentary logs and stratigraphic columns differ in their purposes. Sedimentary logs are detailed graphic displays of the lithologies, sedimentary structures and the fauna of a succession. The succession is broken down into homogeneous units termed *sedimentary facies*, which contain distinctive combinations of features. The manner of deposition of a unit can be inferred from its facies, and the overall environment of deposition from its vertical and lateral variation. There are a number of conventions in recording logs. As with stratigraphic sections, the thickness of beds is shown to scale in a vertical column. However, in a sedimentary log there is also a horizontal scale: the width of the column is a measure of the grain size of the rock unit portrayed (Figure 7.2). Symbols are used to indicate a wide variety of sedimentary features, such as different forms of ripples, cross-bedding, rootlets and mud flakes. So far no convention of symbols has been universally accepted. Devise your own in a form that makes them easy to understand.

Choose a site for a sedimentary log as for a stratigraphic section. Measure the thickness of each lithological unit and record its sedimentological features in your notebook. Take special note of the boundaries between units, that is where they are erosive, sharp or gradational, and see whether there are lateral variations. Draw logs before leaving the field so that any gaps in your notebook information can be identified and rectified. Tucker (2003) gives full details.

7.4.4 Way-up of beds

Symbols indicating which way beds 'young' are frequently omitted on maps in strongly folded areas. But there are ways of telling which way up a bed is. *Sedimentological* indications are the most abundant and include cross-bedding, ripple marks, sole marks, graded bedding, down-cutting erosive boundaries, load casts and many others (Appendix C). Palaeontological evidence includes trace fossils, burrows and pipes left by boring animals and corals in their growing position. Many palaeontological pointers to way-up are fairly obvious, but one alone is not always reliable. Look at a number of different indicators before making a decision, because an incorrect determination of way-up will drastically affect your interpretation of the structure. As with all data recorded, express your degree of confidence with your interpretation. The use of the words 'probably' or 'possibly' in your notes is not a sign of weakness. On the contrary, expressing the quality of data increases their value.

In structurally interesting zones, where it may be difficult to tell which way up the beds are, use the 'overturned' dip and strike symbols on your map where you are sure beds are wrong way up and add a dot to the pointer when you know they are right way up (see list of symbols on inside of back cover); uncommitted symbols then indicate lack of evidence either way. Alternatively, where way-up evidence is available, a Y symbol can be used (with the stalk pointing in the younging direction).

Wherever there is evidence of way-up in such areas, note what it is, such as *c.b.* for cross-bedding, *r.m.* for ripple marks, *t.p.* for trumpet pipes and so on; this is all part of your field evidence.

7.4.5 Grain sizes

Many sedimentary rocks can be classified by their grain size. Anything greater than 2 mm is gravel; anything less than 4 μm is mud; what lies between is sand or silt. Each of these groups is subdivided into coarse, medium and fine, and so

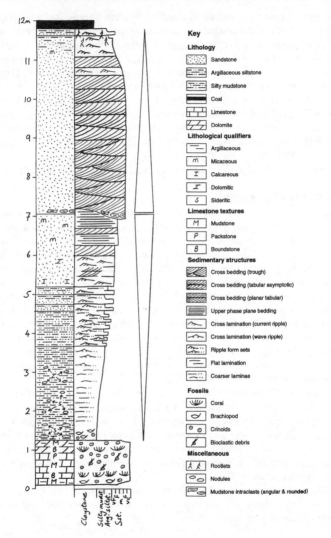

Key

Lithology

Sandstone	
Argillaceous siltstone	
Silty mudstone	
Coal	
Limestone	
Dolomite	

Lithological qualifiers

Argillaceous	
m	Micaceous
I	Calcareous
I	Dolomitic
S	Sideritic

Limestone textures

M	Mudstone
P	Packstone
B	Boundstone

Sedimentary structures

	Cross bedding (trough)
	Cross bedding (tabular asymptotic)
	Cross bedding (planar tabular)
	Upper phase plane bedding
	Cross lamination (current ripple)
	Cross lamination (wave ripple)
	Ripple form sets
	Flat lamination
	Coarser laminae

Fossils

	Coral
	Brachiopod
	Crinoids
	Bioclastic debris

Miscellaneous

	Rootlets
	Nodules
	Mudstone intraclasts (angular & rounded)

Figure 7.2 *A graphic sedimentary log. The horizontal scale is a measure of grain size. Divisions are unequal because the φ-scale range for silt is only four-fifths of that for sand (see Appendix D). The vertical triangles to the right indicate coarsening and fining in the sequence (courtesy of A.R. Gardiner)*

on (Appendix D, Table D.2). Measure larger grains in the field with a transparent plastic scale placed over a freshly broken surface; use a hand lens with the scale for the finer sizes. Generally, if a piece of rock is gritty between your teeth (no need to bite!) then silt is present, and if grains lodge between your teeth there is fine sand, but that should be visible under your hand lens.

7.4.6 Smell

Some sandy rocks also contain clay. Breathe on a fresh surface and note whether it returns a clayey smell. This is not infallible, for if the rock has been too indurated the clay minerals will have been altered to new minerals. Other rocks, namely those that once had a high organic content, emit a sulphurous smell when hit with a hammer. Iron staining indicates an iron cement.

7.4.7 Hardness

Always test a very fine-grained or apparently grainless rock by scraping the point of your hammer across it. If it scratches, it is probably a sedimentary rock, if not it may be a chert or hornfels, or an igneous or pyroclastic rock. Some white, cream or grey rocks can be scratched with your fingernail. They are probably gypsum or anhydrite, or possibly even rock salt, but one lick can settle that!

7.4.8 Acid

Every geologist should carry a small bottle of 10% hydrochloric acid in the field. To use it, break off a fresh piece of rock, blow off any rock dust and add one drop of acid. If the reaction is vigorous, the rock is *limestone*. If it does not fizz, scrape up a small heap of rock powder with your knife and add another drop of acid to it. Gentle reaction indicates *dolomite*. Many carbonate rocks contain both calcite and dolomite, so collect specimens for staining when you return to base. Remember, however, that some rarer carbonates react to acid too. One drop of acid is enough to test for reactions so do not flood the surface with it. All you need is a very small plastic bottle, the type used for eye-drops.

7.5 Fossils

Fossils cannot be considered in isolation from their environment. All the features found in a fossiliferous rock must be recorded if you are to gain the full benefit from the fossil itself.

- Note their abundance in each fossiliferous horizon in the locality.
- Are they widespread or clustered in groups?
- Did the fossils die where they were found or were they transported there after death?

- Do they show alignments due to currents?
- Different fossils may occur in different parts of the same horizon and there may be lateral changes that can be traced over considerable distances, indicating a changing environment. There may also be a vertical change as the depth of water in which the rocks were deposited changed.

All this must be recorded in your notebook, either on a measured section or, if the occurrence is suitable, on a stratigraphic section or graphic log.

Do not be over-anxious to collect a fossil when you find it. First study it in place, noting its attitude and surroundings: make notes and sketches. Probably you will see only a small part of the fossil, perhaps because only a small part of it is exposed, or because only small fragments occur. Decide how best to remove it from the rock, then remove the specimen carefully, trying to keep it intact. Use a chisel or even scrape around it with your knife. Sometimes it is better to remove a large piece of rock and carry it around all day, than to be too ambitious in trying to extract a specimen in the field. If you find a whole fossil, one specimen of that species will probably be enough; leave the rest for others. Usually, however, you will only be able to collect incomplete specimens. Some may show external features, some internal casts. Collect both. As with rocks, name fossils in the field but before going into the field, refer to the types you may expect to see in the rocks you will be looking at. Do not be discouraged if you cannot name in detail every fossil you find. Expert help is often needed.

Once you have discovered a sequence containing specific fossils in some part of your mapping area, you may then find that you can use it for mapping on a wider basis, especially where you have a series of repeated sequences, or cyclothems. The fossils will tell you which part of the series you are in. Again, where you have beds of great thickness, your fossiliferous horizon will tell you where in that bed you are. Mark rich or important fossil localities on your map with a symbol, so that others can find them again later.

7.6 Phaneritic Igneous Rocks

Phaneritic igneous rocks are easily recognised, and acid to intermediate (leucocratic) varieties can usually be readily named. Dark-coloured (melanocratic) phanerites are perhaps a little more difficult to identify, but you can usually put some field name to them that is nearly correct. Before you go into the field, try to look at specimens of the types of rock you expect to encounter – if possible, those from the area you are going to.

7.6.1 Grain-size in phaneritic rocks

Grain-size terminology in igneous rocks differs from that used for sediments, namely:

- Coarse-grained >5 mm
- Medium-grained 1–5 mm
- Fine-grained <1 mm

Use the terms coarse, medium and fine when discussing a rock, but in formal descriptions state grain sizes in millimetres. If a rock is porphyritic (or porphyroblastic if a metamorphic rock), remember to quote the size of the phenocrysts or porphyroblasts too; a phenocryst or porphyroblast 10 mm long may appear to be 'large' in a fine-grained rock, but not in a coarse one.

7.6.2 Igneous mineralogy

When naming a rock, identify the principal minerals and estimate their relative abundances, using the chart in Appendix D. Without a chart, you will almost certainly overestimate the quantity of dark minerals by a factor of up to two. Look at a selection of the grains of every mineral present, not just one or two of them. Identify each mineral in turn, using your hand lens; note the relationships between different minerals. Rotate the specimen in the light to catch reflections from polysynthetic twinning in plagioclase; it is remarkable how many geologists have never recognised this except under a microscope. Dark minerals are the most difficult to identify in a hand specimen, and pyroxene, amphibole, epidote and tourmaline are easily confused. The different cross-sections and cleavages in pyroxene and amphibole should be known to geologists. Note also that the cleavage in amphibole is much better than in pyroxene; epidote has only one cleavage and tourmaline has virtually none. Refer to Dietrich and Skinner (1979) to name the darker rocks.

7.7 Aphanitic Igneous Rocks

Aphanitic igneous rocks can be difficult to name in the field. Hard and compact, at first sight they appear to give little indication of their identity. Divide them into light-coloured aphanites, ranging up to medium red, brown, green and purple; and darker aphanites covering colours up to black. Use the old term *felsite* for the first group and *mafite* for the second. Table 7.1 shows how the two groups divide. Careful examination of aphanites under a hand lens usually gives some pointers to their identity, and many contain phenocrysts, which also helps. Basalt is by far the commonest of all black aphanites. In the field, use the term 'spotted black rock' if all else fails.

7.8 Veins and Pegmatites

Quartz veins are common and should give no problem in identification. Some are deposited by hydrothermal solutions along fractures and may show coarsely

Table 7.1 *Division of aphanitic igneous rocks into felsites and mafites.*

Felsites	Mafites
Rhyolites	Andesites (a few)
Dacites	Basalts
Trachytes	Picrites
Andesites (most)	Tephrites
Phonolites	Basanites
Latites (trachyandesites)	–

zoned structures and, sometimes, crystal lined vugs. Others have been formed by replacement of rock and may even show 'ghosts' of the replaced rock with structures still parallel to those in the walls. Some veins are clearly emplaced on faults and may enclose breccia fragments; some may contain barite and fluorite and even sulphides. If pyrite is present, check for ore minerals (but see Figure 7.3). However, not all veins are quartz veins. Some contain calcite, dolomite, ankerite or siderite, or mixtures of them, and they may be mineralised too. Note, however, that veins do not necessarily have igneous associations.

Pegmatites always have igneous associations. They are usually, but not exclusively, of granitoid composition. Grain sizes may be from 10 mm upwards to over 1 m in size (Figure 7.3). 'Granite pegmatites' fall into two main groups, *simple* and *complex*. Simple pegmatites are usually vein-like bodies consisting of coarse-textured quartz, microcline, albite, muscovite, sometimes biotite and, rarely, hornblende. Complex pegmatites can be huge, some may be tens of metres long, and often pod-like in shape with several distinct zones of different composition around a core of massive quartz. They may be mineralised, with beryl, spodumene, large crystals of commercial muscovite and various other micas, including zinnwaldite and lepidolite. They may also contain ore minerals such as cassiterite and sometimes rarer ore minerals, including columbite (niobite) and tantalite.

7.9 Igneous Rocks in General

Always examine an igneous contact thoroughly. Look at both sides carefully and make sure that it is not an unconformity or a fault contact. Note any alteration and *measure* its extent: a 'narrow' chill zone will mean little to the reader of your report. Sketch contact zones and sample them. Contact metamorphism converts mudstones to hornfelses – hard, dense, fine-grained rocks often spotted with aluminium silicates. They can be difficult to identify; map them as

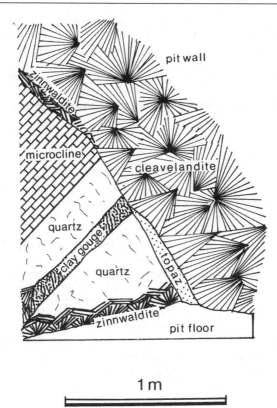

Figure 7.3 *Part of a complex pegmatite in Buganda Province, Uganda, redrawn from a notebook sketch of a prospecting pit. Note the large radiating sheaves of 'cleavelandite' albite, massive microcline, granular topaz (not gem form) and large books of zinnwaldite (a lithium-bearing form of biotite).*

appropriate, for example *grey hornfels* or *spotted black hornfels* or *garnetiferous green hornfels*. Sandstones are metamorphosed to quartzites near contacts. Carbonate rocks become *tactites* or *skarns*, diverse mixtures of silicate minerals. Search skarns with special care for ore minerals, for they are very susceptible to mineralisation. Examine contacts between lavas (and the rocks both above and below them) closely and do not forget the contacts between individual flows.

Few large intrusions are homogeneous, yet many maps give that impression, for they are often shown on a map by only one pattern or colour. Map the interior zones of an intrusion with the same care you would give to an equivalent area of sedimentary rocks. Boundaries between phases may be irregular and gradational, but differences in mineral composition and texture, and very often flow-banding, can usually be seen if looked for. Map them. Map also all dykes and veins in intrusions, or at least note their frequency and strikes if small. Record joint patterns.

7.10 Pyroclastic Rocks

Treat pyroclastic rocks as if they were sedimentary rocks and apply the same rules when mapping them. They are important markers in sedimentary sequences because they may be deposited over wide areas in relatively short periods of time. Pyroclastic materials are essentially glassy ashes. Unconsolidated they are called *tephra*, when consolidated *tuff*. The term *agglomerates* denotes pyroclastics composed mostly of fragments larger than 64 mm; *lapilli tuff* of fragments (usually rounded) of 64 to 2 mm and *ashy tuff* is anything below 2 mm. *Welded tuffs* are those in which the ashy fragments fused during deposition. *Ignimbrite* is a special name reserved *only* for rhyolitic welded tuffs. Name tuffs, where possible, for their related lavas, for example *andesite tuff* or *ashy andesite tuff*, but many fine-grained varieties are difficult to identify in the field and more non-committal names are justifiable. Some tuffs are so glassy, or even apparently flow-banded, that they can be mistaken for lavas in the field. Tuffs tend to devitrify to give spherulitic and perlitic textures. Many weather easily to industrially useful products such as *bentonite* and *perlite*. Thorpe and Brown (1991) give more detailed accounts of how to map igneous rocks.

7.11 Metamorphic Rocks

Contact metamorphism has been dealt with under igneous rocks. Here we are concerned only with rocks resulting from regional metamorphism. Two factors need to be considered when mapping them: the original lithology (protolith) stratigraphy and present lithology. Whenever possible, map them separately.

7.11.1 Naming metamorphic rocks

Fine-grained sedimentary rocks change with increasing metamorphism, first to slates, then to phyllites, schists and gneisses. Igneous rocks deform and recrystallise to gneisses or schists, and many basic igneous rocks, including volcanics, become *amphibolites*.

Name slates for their colour, such as brown, green, grey, blue or purple, and their recognisable minerals, for example *pyritic black slate* or *green chiastolite slate*, and do remember that most slates are not hard roofing-quality slates. Phyllites are more lustrous than slates, with a silky sheen of sericite and chlorite.

Geologists seldom agree where to put the boundary between phyllites and schists in the field: the division tends to be subjective. In general, if individual mica or chlorite flakes can be clearly seen, call it a schist; if not, it is a phyllite. Mica schist is a common 'sack name'. Where possible, define 'mica schists' as muscovite schist, biotite-garnet schist, chlorite schist and so on; but not all schists are micaceous: there are actinolite schists, tremolite schists and many others. Unfortunately, schists tend to weather easily and so are often poorly exposed.

Gneisses are medium (0.73–1 mm) to coarse-grained (from 1 to >2 mm) foliated rocks in which bands and lenses of different composition alternate. Some gneisses split roughly parallel to their foliation owing to the alignment of platy minerals, such as micas; other do not. Always qualify the word gneiss by a compositional name when first used: not all gneisses are granite gneisses as is too often assumed.

Gneisses may also be named for their textures, such as *banded gneiss*. Some may contain apparent phenocrysts or *augen*. They may be cataclased augen, or they may be porphyroblasts of large new crystals growing in the rock, perhaps replacing former augen. You probably cannot tell which until seen in thin section, but augen gneiss is a convenient field name in either case, even if not always strictly correct.

Migmatites are, literally, mixed rocks. They contain mixtures of schistose, gneissose and igneous-looking material. Treat them in the same way as other gneisses: name them for composition, texture and composition. For all these rocks, measure the dips and strikes of foliation and the direction and amount of plunge of the hinges of any minor folds (see Figure 6.13). In many instances, some structure can be discerned by drawing form-lines as described in Section 4.4.2, Figure 4.2 and Marshak and Mitra (1988).

7.11.2 Formations of metamorphic rocks

Just like sedimentary rocks, metamorphic rocks are mapped by dividing them into formations based on diagnostic properties. Some of these properties will relate to the nature of the protolith, others to the character of the metamorphism. Like sedimentary and igneous rocks, formations of metamorphic rocks may have a locality name as a prefix, followed by the dominant lithology, for example Ardrishaig Phyllite Formation. Contacts between metamorphic formations are just as sharp as those between most sedimentary rocks or igneous rocks. Some, however, are gradational, especially within schists and gneisses. Identify every exposure compositionally when mapping them so that gradational boundaries can be inferred where necessary (see Section 4.4.1).

7.11.3 Foliation

Where structure is fairly straightforward, map the cleavage, schistosity or other foliations at much the same density as for sedimentary rocks. If the structure becomes so complex that it is impossible to show it adequately on your map, map it at a larger scale or make numerous sketch maps and notebook diagrams. A map cluttered with tightly crowded clusters of symbols is difficult to interpret by its author, let alone by those who may have to refer to it later on.

In addition to foliation, there are many other structures that need to be mapped in metamorphic rocks. These include the plunge and trend of the hinge lines of any minor folds as well as the strike and dip of their axial planes. Always ask yourself: 'What are the surfaces being folded?' The answer could be bedding, cleavage or other foliations or even ptygmatic veins. This observation is important because it allows dating of the folds relative to other structures. The sense of fold asymmetry is also important. Look for lineations, including intersections of planar features, such as bedding/cleavage, cleavage/cleavage, and so on; or mineral lineations, rodding, mullions and stretched conglomerate cobbles (Figure 6.10). In fact, map any structure, even if you do not know its significance at the time. Its meaning may become clearer later or it may not, but at least you have it on record if it does. For further information refer to Fry (1984) and Lisle (2003).

7.12 Economic Geology

Any geologists worth their salt should be able to recognise the principal economic minerals and rocks, for it is their duty to consider the economic, as well as the purely scientific, aspects of any area they map. To ignore them or to consider them beneath their scientific dignity, as some do and freely admit, is intellectual snobbery.

Before going into the field, review any literature concerning the minerals in the region you are about to map, both metalliferous and industrial. Note records of quarries and mines. Find out what ores have been mined, and particularly, whether they were associated with sulphides, for these ores have distinctive outcrops. Also note the rocks the ores were associated with and keep them in mind when mapping.

7.12.1 Types of body

Ore bodies do not necessarily crop out at the surface in easily recognisable form. Some are just rock in which metallic minerals are disseminated, and often sparsely disseminated at that. Some *stratiform* zinc-lead ores are merely shales with finely dispersed zinc and lead sulphides, similar in grain size to the rock minerals themselves. Porphyry copper deposits, those large stock-like granitoid intrusions that supply more than half the world's copper, often contain less than 1% metal, and look much like any other intrusion. Take nothing for granted.

7.12.2 Oxidation

Ore bodies do not stand up out of the ground with fresh, shining crystals of ore mineral glinting in the sun. Sulphides, in particular, are usually extensively altered above the water table by oxidation. Some oxidise to a highly soluble state (copper, zinc and silver ores are examples) and the metals are leached downwards to be re-deposited near the water-table as a *zone of secondary (supergene) enrichment*, leaving the upper part of the ore body depleted in these metals. Those re-deposited in the oxidised environment just above the water-table form an enriched zone of oxide and carbonate ores, those re-deposited below it are reduced again by reaction with the other sulphides present to form a zone of *secondary sulphide enrichment*. Native metals, such as silver and copper, may also result from secondary enrichment, depending on conditions (Figure 7.4).

		Iron	Copper	Lead	Zinc	Silver	Gold/tin
	ground	Iron accumulates as limonite gossan		Lead carbonates and sulphates present in gossan			Gold and cassiterite occur as minor enrichments in gossan
OXIDIZING	level		Copper minerals oxidize and metal leaches down	No leaching no enrichment	Zinc minerals oxidize and leach down	Silver released from oxidized galena and leached down	No leaching
OXIDIZING	Leached zone	No leaching or enrichment					
OXIDIZING	Secondary oxide zone	iron oxidizes to limonite	Enrichment by malachite, chrysocolla or sometimes native copper	Sulphates and carbonates (anglesirite, cerussite) remain more or less in place	Often massive enrichments of zinc carbonate (smithsonite)	Often major enrichment of horn silver and native silver	No enrichment
OXIDIZING	water						
REDUCING	-table		Enriched by secondary sulphides (bornite, chalcocite, etc.)	No enrichment Galena	No enrichment Sphalerite	Enrichment of native silver and silver sulphide (acanthite)	No enrichment
REDUCING	Secondary sulphide zone	Iron sulphides no enrichment					
REDUCING		Pyrite	Primary sulphides (chalcopyrite, bornite)	Primary sulphides	Silver in galena	Native gold, cassiterite
REDUCING	Primary ore						

***Figure 7.4** Oxidation of sulphide ore deposits, showing how some ore minerals are oxidised and carried downwards; some to re-precipitate above the water-table; some to be reduced again to deposit as new sulphides, or as native metal, just below it, enriching sulphide ores already there. The insoluble iron oxides remain at surface as a mass of rather cellular insoluble iron oxide gossan.*

The iron oxides, invariably associated with sulphide ores, form insoluble oxides that remain at or close to the surface to accumulate during erosion resulting in as mass of cellular limonite called *gossan*. Gossans, and the reddish-brown soils associated with them, form distinctive indications of sulphide mineralisation but that does not necessarily mean that any useful ores are associated with them; only too frequently you find only pyrite below.

Where small groundwater springs near mineralisation have leaked out, rocks may be stained by the brilliantly coloured copper carbonates azurite and malachite, or coated with tiny green crystals of the lead chloro-phosphate *pyromorphite*, so easily mistaken for moss.

7.12.3   Structural control

Pay particular attention to the fracture pattern in any mineralised district, for ore deposition may have been controlled by faults and joints. However, ore may also be controlled by folds, bedding planes, unconformities, lithological changes, and by contacts where granites and diorites have intruded limestones or dolomites. Ore bodies can be of any shape. Some are vein-like, some are irregular masses grading into their host-rocks, some are mineralised rock breccias in collapsed carbonate rock caverns; others are merely an ore-bearing part of an otherwise barren rock, sedimentary, metamorphic or igneous, and these are the most easily missed.

7.12.4   Industrial minerals and rocks

Many of the materials you map have an economic use. The range is extensive and includes rocks themselves, such as limestone (building, cement, chemical neutraliser); marble (building, monuments, etc.); granite (monuments, building facing, ballast); slate and superficial deposits such as gravels and sands (concrete aggregate, moulding sands); clays (bricks, ceramics, fillers). The list is extensive. Note the gravel pits, some of them possibly now lakes, and also the quarries in your area. Build up your background of industrial minerals and refer to Knill (1978), Harbin and Bates (1984) and Evans (1993).

In any study area there will be typical local stones used for everyday buildings and drystone walling – what gives the UK its regional character. More important civic buildings will be likely built of Jurassic limestone (Portland, Bathstone) or granite. There will also be exotic rocks used for decoration, monuments and fascias of buildings such as banks. Guidebooks to local building stones are available for many UK towns and cities and via building conservation websites.

7.12.5   Fuels

Coal is the most likely fuel to be encountered when mapping in Carboniferous areas of the UK, and the coalfields are well documented. Overseas, not all coals are Carboniferous, and in some countries brown coals and lignites are important. Oil shales are gaining importance as a potential energy source for when petroleum sources are depleted and/or the price of liquid and gaseous fuels rises too high to be economic. Hand samples generally contain enough oil to allow them to be set alight. Recent estimates of resources of shale oil, primarily in the USA, Russia and Brazil, exceed the proven conventional oil reserves. Britain has large potential resources associated with the Jurassic Kimmeridge Clay and, in fact, shale oil was produced from a small Carboniferous field in the Midland Valley of Scotland from 1850 until 1964 (Barnes, 1988). Oil shales can be any age from Palaeozoic upwards.

Natural oil and gas are unlikely to be found directly by surface mapping; however, mapping potential oil trap structures such as anticlines located in sedimentary basins is very important. Geologists can also consider if evidence for hydrocarbon source rocks, porous/permeable reservoirs and impermeable seals exists within the geological sequence found in the mapping area. If the area looks like a good hydrocarbon prospect, mapping is usually followed by seismic reflection surveying and deep drilling.

Before embarking on a mapping programme it is well worth carrying out a web and library search for evidence of any historical mining located within your study area.

7.12.6 Water

Water has been described as 'the essential mineral' and geologists in many counties spend a considerable part of their time looking for it. Much of the search for water is geological common sense. Note its occurrence in any area you map and learn from it; the general water profile is shown in Figure 7.5. There are two basic forms of water supply, namely wells and reservoirs.

Shallow wells are common in any country. They are sunk to a water-bearing horizon, sometimes to a weathered rock but often to sands or gravels. The water is raised by a bucket, a hand-pump or a mechanical pump of some sort. They have the drawback of being subject to contamination unless lined to prevent contamination by soil water. Deep wells (i.e. boreholes) are of two general types: those drilled to an aquifer of water-bearing permeable rock, and artesian and sub-artesian wells drilled to an aquifer below a non-permeable *aquiclude* (Figure 7.6). Deep wells are cased, that is lined by pipes to prevent contamination from shallow groundwater. In the UK details of over 100 000 water wells can be found via the British Geological Survey Geoindex website.

Carbonate rocks make excellent aquifers, largely because of their jointing, often enlarged by solution. However, any well-jointed rock can be an aquifer, and in southern Africa Karroo dolerites are an important water source, whilst in East Africa quartzites confined to phyllites serve as aquifers too. Yet again, in the apparently unpromising African basement, granitoid rocks supply water where open-jointed water-bearing zones are sandwiched between tighter-jointed

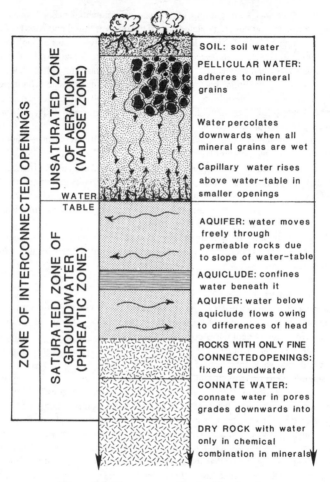

Figure 7.5 *The water profile of soil and rocks.*

unaltered rock below and deeply weathered and partly kaolinised granite above (Barnes, 1988). Do not be too hidebound over your idea of what an aquifer should be.

A reservoir depends on being sited in a valley with catchment that will have a sufficient amount of water to fill it: reservoirs that did not fill are not unknown!

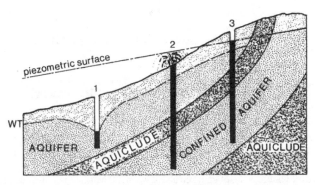

Figure 7.6 *Wells. Well no. 1 penetrates to an aquifer and the water must be pumped. Over-pumping will draw down the rest-level and locally deplete until it is re-charged. Well no. 2 has penetrated through the aquiclude, which confines the water within the aquifer, and because the well-head is below the piezometric surface (the level of the water-table in the aquifer where it crops out), the well is artesian and water flows without pumping. Well-head no. 3, however, is above the piezometric surface and needs to be pumped. Note that the piezometric surface slopes downwards towards the centre of the artesian basin. WT, water-table.*

Geologically, the dam wall must be sited so that it neither leaks around the walls nor beneath it. Cavernous limestones are not good foundation rocks: again, this has been done. There are two types of dam wall: *gravity dams*, held in place by earth- or rock-fill supporting a clay core that prevents the leakage of water through the dam wall or beneath or around the sides; and concrete *arch dams*. Both are matters of engineering geology, beyond our remit here.

8

FIELD MAPS AND FIELD NOTEBOOKS

Field maps and notebooks are valuable documents that constitute part of the record of field evidence on which the interpretation of geology depends. Both are the property of your employers and will be retained by them as part of their permanent records when you leave them. The reason is obvious: if your former employers wish to reinvestigate an area you mapped, then it will be necessary to refer back to the original records. Remember, your employers are the ones who paid for the work to be done. Students are obviously not in the same position as employees but they are often taken aback when, in their first job, their employers demand their field maps and notebooks for their permanent record.

8.1 Field Maps

Geological mapping is like solving a jigsaw puzzle. It is more difficult in the early stages when very little of the puzzle has been completed, but gets progressively easier as parts of the picture start emerging. The field map is used to visually display the data collected to date, as well as to show the interpretations made of those data. It depicts the partly completed puzzle and therefore is an essential aid to planning the next stage of the work and eventually to completing the map. Data recorded only in the field notebook is not going to be much use to the mapper; because the spatial relationship of the already gathered data will not be obvious. For this reason it is important to plot information on the field map as soon as you collect it.

8.1.1 Data needed

A field map is an aid to the systematic collection of geological data in the field and shows the evidence on which the interpretation of the geology was made. It shows the geological features you actually see in the field; it also shows geology you have inferred from indirect evidence, such as changes in topography or vegetation, spring lines or float. If possible, all contacts should be plotted on it when you are in the field, though some may be inferred from minor indirect evidence or sometimes merely by your judgement of where they most probably occur, helped by interpretation from aerial photographs and by

Basic Geological Mapping, Fifth Edition.
Richard J. Lisle, Peter J. Brabham and John W. Barnes.
© 2011 John Wiley & Sons, Ltd. Published 2011 by John Wiley & Sons, Ltd.

interpolating structure contours back in the camp. However, *fact must always be clearly distinguished from inference*.

A field map is not merely a rough worksheet on which temporarily to plot information before transferring it onto a 'fair copy' map back at camp or base; it is a valuable research document that you or others may wish to refer to again at some later date. No evidence should be erased from it to 'tidy it up' or because it is not needed to aid the present interpretation, nor should you add anything to it at a later date that you think you saw in the field but did not record at the time. DO NOT TRANSFER YOUR DATA FROM THE MAP USED IN THE FIELD TO A NEATER CLEANER 'FIELD' MAP IN THE EVENING.

The type of information to be recorded on a field map is:

1. The location of all rock exposures examined.
2. Brief notes on the rocks seen.
3. Structural symbols and measurements, such as those for dip and strike.
4. Locations to which more detailed notes in your notebook refer.
5. The location from which each rock or fossil specimen was collected.
6. The location at which every photograph was taken or field sketch made.
7. Topographic features from which geology may be indirectly inferred but which are not already printed on the map; examples include changes of slope or vegetation and the position of seeps and spring-lines.
8. All contacts, including faults, both certain and inferred.
9. River terraces, beach terraces and other similar features.
10. Alluvium, scree, boulder clay and any other superficial materials, including landslide debris.
11. Cuttings, quarries and other man-made excavations exposing geology, for example pits and boreholes; even mine spoil heaps can indicate geology beneath the surface.
12. Comments on the degree of exposure or lack of exposure, and on soil cover.

Because they are valuable, field maps should be kept clean and protected from rain and damp. This is not always possible and important information must not remain unplotted for fear that the map may get wet or dirty if the map case is opened in the rain.

Ink in your day's work in the evening, including your map notes written in small but legible printing, using a fine pen and *waterproof* ink.

8.1.2 Preparation

Some base maps are so crowded with topographic information, such as heavy contour lines, that this is likely to overwhelm any geological information later plotted on them. In such cases, experiment with lightened photocopies of the map before departing for the field.

Before using a new map sheet, cut it carefully into a number of like-sized rectangular parts, or 'field slips', to fit into your mapping case without having to fold them. Folding ruins a map: it is difficult to plot any information close to creases (especially if folded over twice) and any information plotted close to the crease, or on it, is soon smeared and eventually worn off. Before setting off for the field check that your field slips fit together with gaps or overlaps. Any gaps would be a disaster, though overlapping slips can be irritating too. On the reverse of every field slip print the title of the (complete) map, the number of the field slip and a diagram to show how the different field slips should be assembled. This is in case field slips are mislaid in the headquarters filing system. Also on the reverse of at least one slip, state the scale and a full explanation of the colours and symbols used (Figure 8.1). Any unconventional colours or symbols used on a single slip should be shown and explained on the back of that slip. On the face side of the map, the north direction from which all readings have been measured should be shown – true, grid or magnetic as the case may be. In addition, the slip should show its author's name, the dates of the fieldwork period and the notebook numbers relating to it.

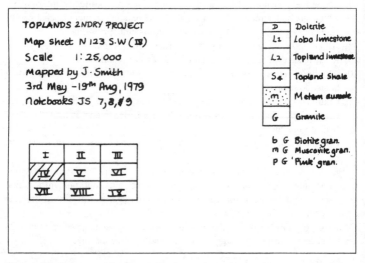

Figure 8.1 *Reverse side of a field slip with the information it should carry. Note the index showing how this slip relates to those that make up the whole mapping area.*

Generally, it is better not to stick your field slips together when fieldwork is complete. It makes the map awkward to use again in the field if new information is to be added, and also most adhesive tapes shrink and dry out with time.

8.1.3 How and what to plot

A field map is a record of field observations of the type listed in Section 8.1.1. Plot the position of exposures seen and indicate rock-type by formation letters, letter symbols or by colouring, supported where necessary by notes on the rock condition. Keep notes short and use abbreviations such as: *fg* (fine-grained), *lam* (laminated), *shd* (sheared). Refer to Section 4.4.1. Many exposures need no more than an outline to show their limits, shaded with the appropriate coloured pencil. Exposures that are so small that they can be shown with a dot should, however, always be supported by a letter symbol, otherwise they tend to be overlooked when inking in the map later. If notebook notes are made at an exposure, then the exposure's location on the map must be linked to the notebook record (see Section 8.1.5). Structural observations are shown by the appropriate symbols, drawn large enough to enable them to be traced off accurately onto the fair copy map on a light table: 6–7 mm is a suitable length for strike symbols on a field map. Print the numerical values of dip or plunge legibly in such a position that there is no ambiguity over which symbol the figure refers to. Even better, record both strike and dip together (Figure 8.2).

Inferred contacts are shown by broken lines, and different reliabilities of inferred contacts are shown by the frequency of the breaks. Figure 8.3 suggests line styles corresponding to the width of the unexposed ground *on the map*. Do not try to distinguish between different types of contact by different line thicknesses. Keep breaks in broken ('pecked') lines small (1 mm), otherwise the lines look untidy. Distinguish faults by '*f*' or, if the dip is known, by a dip arrow (see list of symbols inside back cover).

Show thrusts with the traditional 'saw teeth' on the *upper plate* (*hanging wall*), but do not try to draw the teeth as closely as those you see on printed maps: a tooth every 1–2 cm is quite adequate on a field map, and if you do make a mistake is far easier to erase after inking in (Figure 8.2).

Although a field slip shows factual data this is not its exclusive use. The field map is used for plotting the inferred positions of contacts deduced from indirect evidence such as vegetation, spring-lines and breaks of slope. In fact, the field is the proper place to infer contacts, for there is usually some evidence, however slim, of their positions. The drawing of contacts in the office, or back in camp, is only justified if you have to resort to geometrical constructions such as structure contours, where there is a lack of evidence on the ground, or when you have geophysical or photogeological information to help you.

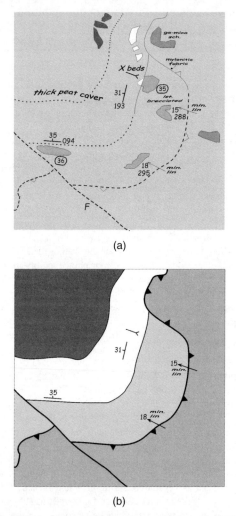

(a)

(b)

Figure 8.2 Symbols for field and fair-copy maps compared. (a) Strike and dip, plunge and azimuth are shown on the field map, though the quadrant of the dip is omitted because this is clear from the tick on the strike symbol. (b) Only dip and plunge are shown on the fair copy.

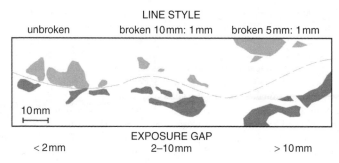

LINE STYLE

unbroken   broken 10mm: 1mm   broken 5mm: 1mm

10mm

EXPOSURE GAP
< 2mm   2–10mm   > 10mm

Figure 8.3 *Styles of line used for contacts. The type of line used depends on the degree of certainty of the position of a contact.*

Any topographical features that may reflect concealed geology but are not already printed on your base map should be added to it; these include: vegetation changes, distinctive soils, springs, swampy patches and breaks-of-slope (see Figure 4.3). Show also landslides, scree and alluvial terraces. Outline mine tips because they can often provide fresh specimens of materials that are otherwise seen at surface only in heavily weathered state (this is particularly so in humid tropical climates) or even prove that certain rocks that are not exposed, or are quite unsuspected in the area, do not occur at depth.

Of course, the recorded strike of dipping units provides valuable clues for inferring the direction of contacts. However, this does not mean that you simply take a ruler and draw a straight contact parallel to the measured strike. In areas of rugged topography use the method of structure contours to predict outcrop patterns (see Section 4.5.4). This will lead to directions of contacts that deviate from that of strike, and give vee-shaped contacts in valleys and across ridges.

If space on the field slip permits it, write neat brief notes to record the nature of evidence used in an interpretation of a contact, for example 'sink holes' or 'breccias'. But the amount of detail that can be shown on a map obviously depends on the scale you are working on. A field slip should not be so cluttered with information that the 'wood cannot be seen for the trees', but even more difficult to interpret is the map that shows almost nothing but a series of numbers referring to notebook entries. There is a happy medium between these two extremes. The face of the map should contain all relevant *basic* geological information; the notebook should expand on it and provide details of features too small to show on the map. The complexity of the geology and its degree of exposure also, of course, determine how much can be shown. However, if the main object of the work is to solve a specific geological problem, then the scale must be large enough to show, without crowding, the type of detail that must be mapped to solve the problem, and if a map of suitable scale is not available, then one must be made. Often, the complexity of geology and degree of exposure vary from one part of a region to another, so that very large-scale maps need to be made only over limited areas, with a considerable saving in cost. Frequently, the results of small-scale mapping indicate areas that require re-mapping on a larger scale; this is particularly so in mineral exploration where maps of larger and larger scale but smaller and smaller areas may be made as the more interesting localities are recognised and unmineralised ground is eliminated. If *occasionally* you are forced to make more extensive annotations on a map than space will allow then make a small needle hole at the locality, write *PTO* against it and write your notes on the back of the map, if not already written on, but do not make a general practice of it. If you need more space, use a larger scale or make a detailed map of a limited area as described in Section 4.9.

8.1.4 Neatness

Information written on the map in the field must be written as legibly as circumstances allow. Keep one pencil, a reasonably hard one, for plotting on your map, and another for the notebook. Keep your plotting pencil needle sharp, otherwise you cannot write legibly on your map. If you use the same pencil for map and notebook, you will be constantly sharpening it between note taking and plotting. Do not use ballpoint or felt-tipped pens. Write on your map in a fine, clear printed script. Do not use miniature cursive 'joined-up' handwriting; it is far less legible, especially when written with ice-cold hands under the often difficult conditions of fieldwork. Do not use stylus-style pens in the field; everyone makes mistakes and they are far more difficult to remedy if made in waterproof ink; secondly, notes frequently have to be erased and rewritten because they overlap some geological feature you had not found when you first wrote them. Even when inking-in pencil-written notes, you frequently have to rearrange them so that they are neater, more legible and parallel to one another. Drawing and draughtsmanship are essential skills for a geologist; if you cannot draw neatly, you cannot map accurately. Much of this skill can be acquired by effort and practice.

The map published by Ramsay and Huber (1987, p. 683) serves to remind us of some attributes of a good field slip:

1. A clear distinction between observation and interpretation, in terms of colouring and the style of drawn boundaries.
2. A neat presentation that allows a wealth of detail to be shown legibly on the map.

3. Written notes where space allows, such as in the lochs, which helps make the map more 'self-contained'.
4. A style of interpretation that appears compatible with the types of structures seen at outcrop scale.

8.1.5   Linkage of map localities to notebooks

The most practical permanent way to link map localities to notebook notes is to use map (grid) references (Section 3.3.2). Map references have the advantage that points can be located by a group of figures with great accuracy and without ambiguity and, even if the original field slip is lost, the points can be relocated on any base map of any scale covering that area. In general, however, re-finding points on your map using map references during interpretation is slow and irritating. Easier to use is a simple consecutive numbering of observations. This works well providing the points on the map are fairly closely spaced along more or less specific directions, such as traverse lines. Consecutive numbering is improved by designating each grid square printed on the map by a letter, or by the map reference of the SW corner of each grid square, and then giving consecutive numbers to the localities visited within those squares. Whichever you do, always enclose locality numbers written on the map within a circle to avoid confusing them with structural readings. Notebook entries are numbered A1, A2, A3 ... or 8746/1, 8746/2, 8746/3 ... and so on (Figure 3.2) and there is seldom any difficulty of relocating them on the map, but always draw a diagram in the front of your notebook to show later readers how the letter symbols relate to map squares. The drawback is that if you do lose the field map the notebook becomes virtually useless.

Map localities can also be identified by notebook page numbers. If several notes are made on the same page, designate them a, b, c, etc. When more than one notebook is used for a project, prefix the page number with the book number: locality 5/23b, for example, means note b on page 23 of notebook number 5. Much depends on whether you are mapping purely for your own purposes, such as on a training course or for your own research project, or whether you are working for an organisation. Organisations usually have their own rules so that later workers can use your field maps and notes for later reassessment of a region.

8.1.6   Inking and colouring field slips

Observations made on field slips during the day should be inked-in that evening. Even on the best protected maps, fine pencil lines become blurred or lost with time. When green-line mapping, exposure limits should be outlined with green *waterproof* ink or, in sunnier climates, with a fine, closely dotted (not dashed) line in black waterproof ink. After inking, re-colour each exposure with the appropriate coloured pencil. Ink the traverse lines with a continuous line where geology is exposed or certain, and a broken line where inferred, then overlay the traverse line, continuous or broken, with a pencil line of appropriate colour.

Inked contacts can now be distinguished by lines of different thickness drawn with stylus-type pens, to distinguish faults and unconformities from other formation contacts, but notes should still be added to confirm their characters: abbreviations such as '*f*' for fault and '*uc*' for unconformity are sufficient. Unmarked contacts are assumed to be normal.

Ink-in all structural symbols and rewrite the values of strike and dip, and so on. Rewrite notes in fine neat script so that they cover no geological features, and align them so that all, as far as possible, are parallel with the same direction. It is irritating having to turn a field slip first one way, then another, to read the information on it.

There may well be interpretative lines shown on a field slip at the end of a day that are still uncertain: leave them uninked until you can confirm their validity, even if it means re-pencilling their traces each evening to avoid losing them. Add any information from aerial photographs to your map in waterproof ink (purple for general geology, red for faults) to distinguish this information from that gathered on the ground. This discrimination in no way diminishes the validity of photogeological information; but it does distinguish its sources and also indicates where features should be sought on the ground.

Having inked-in your map and reviewed your day's work, then, if you must, *lightly* shade or cross-hatch those areas that you now infer are underlain by specific formations. Do not colour in your map heavily, as if it were a final 'fair cop' map. Do, however, re-colour your traverse lines and areas of exposure more strongly, so that they stand out as the evidence that justifies your interpretation. Geologists only too often map exposure-by-exposure during the day, carefully distinguishing what they have observed from what they have inferred only to obliterate all their field evidence in the evening by swamping the map with solid colour in an endeavour to make it look like a published geological map. A field map is an 'evidence' map; it is not a rougher version of a fair copy, and it should not look like one. Make sure that fact can be distinguished from inference on it.

8.2 Field Notebooks

Like field maps, field notebooks are valuable documents that form part of the record of field evidence on which the interpretation of geology depends. A field notebook will be referred to by later workers reinvestigating the area at least as often as they refer to the field map it relates to – perhaps to elucidate data on the map, perhaps to obtain details of specimens or fossils collected. Later workers may also want further details of specific exposures or lithological sections, to discover why you drew the conclusion you did. Alternatively, your notebooks may provide information that is no longer available: exposures may have been built on or dug away, pits and quarries may have been filled, or their records lost or destroyed. Notebooks must therefore be kept in a manner that others can understand and, above all, they must be legible. The US Geological Survey insists that notebooks are written in hand-printed script. This helps to make even those notes written with ice-cold hands on a wet and windy day more legible. Sketches and diagrams too must be properly drawn and labelled, dimensions given and, where appropriate, tinted by coloured pencil.

Develop a habit of using your notebook. During a project, non-geological records need to be kept too, such as expenses; where better than in the back of your field notebook? Use your notebook as a diary; even if no work is done on a particular day, record the reasons in your notebook. Only too often notebooks suffer the fate of new diaries: copious neat notes are written on the first day but by the end of the week notes are sketchy, untidy and illegible. Bear in mind that the notes you make will probably form the basis of a report written after the fieldwork is completed; make sure they are detailed enough for that purpose. Your field notebook is as important as your field map. Use it properly.

8.2.1 Preliminaries

A well-organised notebook will help you and others retrieve information a later stage. Here are some suggestions:

1. Write the name of the project, the year and the notebook number on the cover of every book. Inside the cover write your own name and address in waterproof ink, and offer a reward for its return if found. Be generous: the loss of your notebook could be disastrous. If necessary, repeat the information in the language, and the calligraphy, of the country you are working in.

2. Number notebook pages but leave the first few free for an index to be filled in day by day, with the date, what was done and the page numbers. This helps not only others who have to use your notebook, but also yourself when you come to look for information when later writing your report from notes made weeks, sometimes months previously. Remember that you keep a notebook to refer to, so make it easy to do so. Ask yourself 'What use would this notebook be if I had to refer to it again in a year, two years, even five years from now?' If you doubt that you could understand it yourself, no one else will be able to do so.

3. Start each day's notes on a new page, stating the date, the weather conditions and the plan for the day's work.

4. Add also to your notebook registers of rock specimens and fossil specimens collected, and photographs taken (see Section 6.11.3). Tape on to the last

few pages photocopied tables of pace lengths, and charts such as those in the appendices. Also fasten a small piece of fine sandpaper onto the inside of the back cover, to keep your pencils sharp.

8.2.2 Linking notes to map localities

Methods to link observations made on your map to notes in your notebook, and vice versa, are given in Section 8.1.5. Write the map references or locality numbers in a column on the left-hand side of the notebook page (Figure 8.4). Use this column only for locality notes, specimen and photograph numbers. Write notes in pencil; specimen and photo numbers in red and blue coloured pencil, respectively, so that they can be quickly spotted. If numerous, give specimens and photos a column of their own.

8.2.3 Recording information

1. The purpose of a field notebook is to expand information from your field map, not to duplicate it. For instance, normally there is little point in writing down the values of strikes and dips plotted on the map unless weather conditions are so bad that you cannot plot them in the field (although some geologists have other views). If for any reason strikes and dips have to be recorded – joints, for example – then make the information easier to retrieve by recording it on the right-hand side of the page.
2. Make your notes as brief as possible, even omitting verbs at times, provided that you do not lose the sense or meaning. Use abbreviations where appropriate; there are many all geologists understand, for example *lst* for limestone, *sst* for sandstone, *sch* for schist; *flt* for fault and *jt* for joint. Tabulate in the front of your book any non-standard abbreviations you use unless their meanings are obvious. Use any short cuts in the field that save time without loss of information. Avoid woolly descriptors such as 'fairly large', 'moderately steep'; use quantitative measurements.
3. Some geologists use separate columns in the notebook for raw observations and interpretations. Whether this practice is adopted or not, it is important to distinguish these in your note-taking.

Whether you ink-in your day's notes is up to you. This author (RJL) has never done so. Figures 8.4 and 8.5 are examples of pages from field notebooks.

8.2.4 Sketches and detailed maps

For many types of field observation a totally verbal record is inadequate. For example, geometrical features resulting from tectonic, sedimentary or magmatic processes are best described with the help of a field sketch. Use sketches to supplement notebook descriptions whenever possible. Be generous to the space

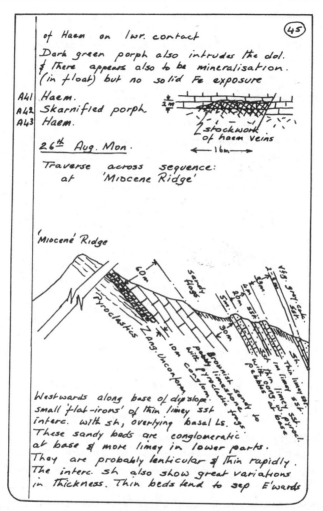

Figure 8.4 Page from a field notebook (John Barnes). The column on the left shows the registered number of specimens collected. The observation number referring to the locality (Ravange, Iran) appears in the same column on the previous notebook page. The cross-section on the lower part of the page needs no locality/observation number as the section line is shown on the field slip and on the map.

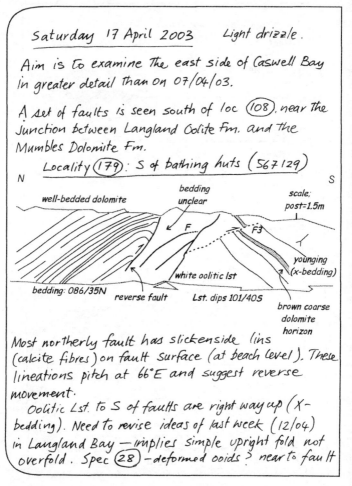

Saturday 17 April 2003   Light drizzle.

Aim is to examine the east side of Caswell Bay
in greater detail than on 07/04/03.

A set of faults is seen south of loc (108), near the
junction between Langland Oolite Fm. and the
Mumbles Dolomite Fm.

Locality (179): S of bathing huts (567 129)

N   S

well-bedded dolomite   bedding   scale;
                              unclear   post=1.5m

                                        F   F3

                                                    younging
                                                    (x-bedding)

            white oolitic lst

bedding: 086/35N   reverse fault   Lst. dips 101/40S

                                              brown coarse
                                              dolomite
                                              horizon

Most northerly fault has slickenside lins
(calcite fibres) on fault surface (at beach level). These
lineations pitch at 66°E and suggest reverse
movement.
     Oolitic Lst. to S of faults are right way up (x-
bedding). Need to revise ideas of last week (12/04)
in Langland Bay — implies simple upright fold not
overfold. Spec (28) - deformed ooids? near to fault

Figure 8.5 *Another example notebook page (RJL). Note the comment on the weather and the reason for the re-examination. There is both a locality number (179), and a grid reference (567 129) given, plus a description, that is, 'S of bathing huts'. The cross-section has been cleaned up in the office later for inclusion in a report.*

allotted to a sketch; use a full page or double page. Remember to leave space around a sketch for labelling.

Geological sketches are quite different from the representation of an exposure produced by an artist. The geologist filters out some information in the sketch in favour of salient geological features. Depending on the nature of the mapping project in hand, details of vegetation, loose debris and superficial coloration patterns due to weathering, algal growth, and so on, on the rock surface may be ignored. Also light and dark tones due to lighting conditions on the irregular exposure surface will be filtered out. Instead, the emphasis is placed on tracing the contacts of intrusion, the form of bedding planes, sedimentary structures and faults. These are drawn as clean continuous lines rather than multiple hatched line work preferred by the impressionist artist. Use a hard pencil (H or 2H). For examples, of good field sketches, see Moseley (1981) and Coe *et al.* (2010).

As with the field map, interpretation such as the extrapolation of a fault line through poorly exposed parts of the locality can be attempted, but remember to use a dashed line to make it clear that this part was not directly observed.

Sketches should show dimensions, or at the very least, some indication of scale. In your drawing, try to maintain the same scale horizontally and vertically. Estimate the height of cliffs before drawing them. First, step back from the outcrop and sketch the basic outline, for example skyline, beach level, major features; then fill in detail. This will help to maintain features in their true proportion. It is also essential to record the orientation of the exposure. For sketches of steep exposure surfaces, the best way to do this is to label the left and right directions as compass points, for example by writing NE and SW on the top left and top right corners of the page, respectively. The orientation of sketches of near-horizontal surfaces, such as wave-cut platforms, can be shown by a north symbol.

The structures and lithologies depicted in the sketch should be labelled; where space is limited, such annotation can be done by using a curved arrow to link the notes to the relevant part of the sketch. The sketch should be given a title. The clarity of a sketch can be improved by shading or lightly colouring different rock units. Figure 8.6 show examples of good and poor field sketches.

Remember the acronym OASIS for sketches

- Orientation
- Annotation (labelling)
- Scale
- Interpretation/Importance of what you are sketching
- Simplify

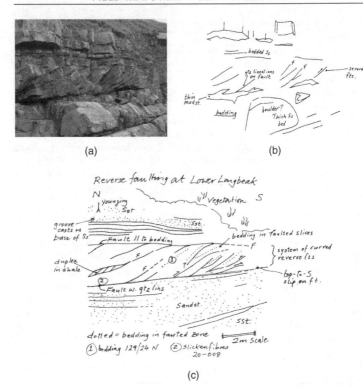

Figure 8.6 Poor and good sketches. (a) Photo of exposure of thrusted sandstones and shales, N of Widemouth Bay, Devon. (b) Poor sketch with no scale, orientation, indefinite linework, and lack of labelling. (c) Good sketch with attempt to trace individual bedding planes and faults as continuous features and representing cross-cutting relationships.

Some exposures may display small-scale features that are impossible to represent on the scale of the field slip. Using the methods described in Chapter 3, construct a larger-scale map of the exposure on a single or double page of your notebook. Such a map should have a north symbol and scale bar.

8.2.5 Sketch cross-sections

An important use of the field notebook is as a scientific diary. In addition to being a repository of observational data, it can be used to record the development of your ideas during the course of the mapping. Take some time out from your outcrop visits to sit down, preferably at a good vantage point with panoramic views, and collect your thoughts about the geology of the area. Use the notebook to summarise your current thinking on the stratigraphic succession, the geological history and the geological structure. For the latter, drawing a sketch cross-section in your notebook is essential. Devote a full page to the sketch, and consider what the map pattern, bedding dips and way-up data imply about the large-scale structure. You may have alternative ideas of the structure, so draw sketches of the different possibilities. The reader of the notebook will be impressed by your consideration of alternative hypotheses – a sign of sound scientific practice. At regular intervals throughout the fieldwork, draw new cross-sections incorporating your new data and your latest ideas. In your notes refer back to your earlier attempts and mention the aspects of the interpretation that have changed. In this manner the notebook is used for 'talking to yourself' during the project.

These cross-sections need not be constructed with great accuracy; they may be more conceptual than exact. Nevertheless the thought processes involved in their construction will force you to think in 3D during the mapping. They should highlight parts of the area where data are lacking and thus may suggest key areas where the effort of the remaining field days could be most spent most profitably.

'ODIN'

Odin, the name of a god in Norse mythology, reminds us of the essentials of a field notebook:

Organisation (e.g. links to map, clear page layout, page numbers, spaces between locality descriptions, dates, separation of raw observations and interpretation, a column for sample, photo numbers).
Detail (data-rich notebook, mini-maps, sketches).
Illustrations (e.g. ample sketches, enlarged-scale maps, cross-sections, schematic diagrams to convey ideas).
Neatness (e.g. legible handwriting, carefully drawn sketches).

9

FAIR COPY MAPS AND OTHER ILLUSTRATIONS

9.1 Fair Copy Maps

Geological field maps are records of factual observations made in the field. When mapping has been completed you must compile a fair copy map interpreted from your field map, your notebook notes, your follow-up laboratory work on rock and fossil specimens, and your library research. This fair copy may be a hand-drawn 'manuscript' map, or it may be aided and drawn by a computer program. The advantage of a computer-drafted map is that it can be easily modified and redrawn to accommodate new data. However, every geologist should be capable of producing a manuscript map, and the following describes the process.

A fair copy map is not merely a redrawn version of the field map; geological formations are now shown as continuous units instead of disconnected exposures. It is also a selective map, and it may well be that some formations distinguished on the field slips are no longer differentiated when transferred to the fair copy. This may be because distinctions made in the field were found to be geologically less important than first thought, because they were so discontinuous that they could only be traced over short distances, or because your employer's policy is to map geology at a larger scale than that intended for publication.

Alluvium, swamp, peat and bog are shown on the fair copy, as are also laterite and boulder clay, but soils are not. The general rule is to show any features that add to the understanding of the geology and geological history, and to omit those that do not.

Much of the information gathered during fieldwork is not transferred to the fair copy. For instance, the notes made on the face of the field map are not normally shown on the fair copy, although generalised comments may be made, such as 'cordierite schist' if this zone is not represented by a specific colour, pattern or symbol of its own, or 'red soils' to justify the continuation of an unexposed dolerite dyke. More specific notes, such as 'malachite stains', may on occasion be needed, but otherwise notes are usually only made to emphasise specific or unusual details, or to justify the geology. Specimen locality numbers should not be shown on the fair copy although rich fossil or ore mineral localities

Basic Geological Mapping, Fifth Edition.
Richard J. Lisle, Peter J. Brabham and John W. Barnes.
© 2011 John Wiley & Sons, Ltd. Published 2011 by John Wiley & Sons, Ltd.

may be marked where their presence is geologically or economically significant. The criterion of what to show is mainly a matter of common sense. The finished fair copy should show the geology of the region in such a manner that the geological formations can clearly be distinguished one from the other and, if they are continuous units, it should be possible to trace them from place to place across the map even though poorly exposed on the ground. Structural symbols should be sufficiently clear that the sequence of events can be elucidated and the stratigraphy determined. The fair copy map allows interpreted large-scale features to be shown such as the axial traces of major folds, and the outcrop of large faults. Data from small-scale features should also be shown but in places some symbols may need to be omitted for reasons of limited space. Above all, the map should be neatly drawn, the colours smooth and distinctive and the printing legible.

It may be asked why so much information, painstakingly collected in the field, is omitted from the fair copy. It is because the fair copy is only a part of your interpretation. It is essentially an index that provides the basis for understanding any accompanying report. The map is not an end in itself, but it should still be able to stand on its own, showing the general features of the geology in a clean and concise manner.

9.2 Transferring Topography

A fair copy map is usually drawn on a fresh copy of the original topographic base map used in the field. If possible, the fair copy map should be produced as a single sheet. If for the lack of an original base map, or because the only base map you had was heavily cluttered with coloured geographic information or with colour layered contours, the fair copy will have to be made on tracing film, then geographic grid lines should be drawn. In addition, sufficient topographic contours must be traced off it to make the geology understandable. This is tedious but necessary. In very mountainous districts, sufficient relief may sometimes be shown by tracing off every second or even fifth contour. Draw the contours in brown so that they can be distinguished clearly from geological boundaries.

If you have to make your base map from aerial photos, as one sometimes must, trace off the main drainage and hill tops to at least give some geographic frame for the geology.

9.3 Transferring Geology

When preparing a fair copy map, information has to be transferred from the field slips onto a clean base map. There are several ways of doing this. The first is to copy from field slip to the fair copy purely by inspection; this can only be done where there is sufficient printed detail on the map to serve as reference points. If there is not enough background, then divide the grid squares into smaller squares on both slips and base map, and transfer information square by square; in both cases, strike and dip symbols must be re-plotted from the original data. A better way is to use a *light table*, so that detail can be traced directly from field slips to fair copy, even if the fair copy is a non-transparent paper map. If the printed geographic detail on the well-used field slips does not fit that on the brand new fair copy base map, do not be surprised; this is due to shrinkage caused by weathering and is probably slightly different across the paper than it is up and down it, a result of how the paper was made. If grid squares are not already on both slips and base map, draw your own grid, and adjust each field slip to the best fit as you trace the geology, square by square.

9.4   Lettering and Symbols

Bad printing can ruin an otherwise well-drawn map whilst good lettering can often improve a poorly drafted one. Producing your map by using a computer drawing package has major benefits in this respect. For hand-drawn maps, 'transfer lettering' is one solution, but it is expensive. Many organisations have machines to print map lettering of different styles and sizes onto transparent adhesive tape, which can be stuck down on the map (after colouring). Stencils can also be used in default of any other method, and although they do not give such a good result as printed lettering, they can be used over and over again, but mistakes are difficult to remedy. Alternatively, you can also print your lettering on a computer and printer, using different styles and sizes, cut them out and stick them down with paste. It all depends on what your department or organisation has available.

Despite such aids, every geologist should develop a good legible hand-lettering style, for there will be many occasions when this is the only way to add text to a map; inking-in your field map is a good way to practise. *Italic* letters are easier to print than upright, and always use parallel guide lines, except for the very smallest lettering, to control the regularity in size of lettering of different import.

Draw all strike symbols on a fair copy exactly the same size: 5 mm long is suggested. Lineation arrows can be a little longer. Draw arrowheads neatly with an ordinary mapping pen. Print figures for dip (bearing for strike is now omitted) parallel to the symbol, or parallel to the bottom edge of the map, but not in both directions in different parts of the same map (see Figure 8.2). The symbols printed inside the back cover of this book conform to those generally accepted around the world. A much more extensive list compiled by the staff of the Australian Bureau of Mineral Resources is given by Berkman (2001).

9.5 Formation Letters

Every rock unit that appears on the face of the map, whether sedimentary, metamorphic or igneous, must have a distinguishing symbol or 'formation letter(s)' assigned to it. Established formations may already have officially recognised symbols such as d6a for the Lower Pennant Measures of the British Carboniferous System, M1d for the Mississippian Leadville Dolomite (Formation) in the USA, and M2 for the Upper Red Formation, better known as the 'Fars', of Iran. If no symbol has been allocated to a formation, you could do it yourself, following the convention of the country, if there is one. Otherwise, use the initial of the unit where possible, as this acts as a mnemonic. Avoid calling your units A, B, C or 1, 2, 3. Show formation letters on every area of each rock unit that appears on the map. Where a unit covers a very wide area, repeat the formation letters in several places, but where an area is too small to contain them, print the letter beside it with a 'leader' (zig-zag) pointing to it.

9.6 Layout

A fair copy map should be properly laid out. It should have a proper title, a scale, north symbols (true, grid) and an explanation of the other symbols used, together with a record of the authorship(s) of the map and an accreditation of any sources used, including the source of the base map itself. It should also show the dates the fieldwork started and finished and the date of publication. The arrangement of this matter requires some thought and may warrant making a mock-up on tracing paper or, at the very least, a rough sketch so that the sheet looks properly balanced; if the map is stuck down on a larger sheet of good quality paper, the ancillary information can be arranged around it, or at one side of it, as in Figure 9.1. Express the scale of the map in figures as a ratio or 'representative fraction' (e.g. 1:15 000) and graphically as a scale bar (Figure 9.2). The latter is useful if the map ever gets copied at an altered scale. North symbols should be as plain as possible, so that they are unambiguous; avoid those fancy north points that draftsmen love to draw if left to themselves; exaggerate the differences of the angles of grid and magnetic north from true north so that there can be no mistake of their relative positions. Beside them print the amounts by which they deviate from true north, and do not forget to give the annual rate of magnetic change. In the explanations, draw the symbols at exactly the same size as you used on the face of the map.

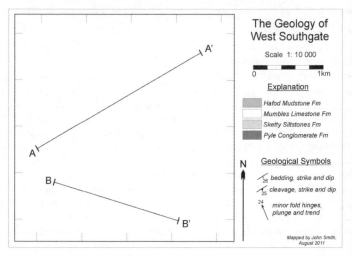

Figure 9.1 *Layout of a fair copy map, showing the arrangement of explanatory matter. Note the terminations to the cross-section lines so there is no ambiguity over the exact positions where the section lines end.*

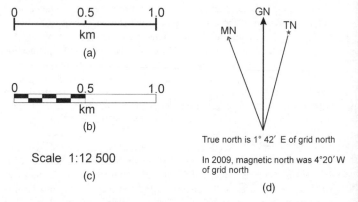

Figure 9.2 *Scales and north symbols for fair copy maps. (a,b) Graphic scales. (c) Scale expressed as a representative fraction. (d) Composite north symbol shows the declination and grid magnetic angles diagrammatically.*

9.7 Colouring

A manuscript fair copy map is normally hand-coloured by its author. Colour your map by whatever method you think you are capable of doing neatly, and which the paper will take. Many maps are spoilt by this final colouring. Watercolours probably give the more finished look to a map but they are difficult to apply over large areas, especially if those areas have intricate boundaries: irregular drying marks spoil the result. Colouring pencils give excellent results with a minimum of practice. Lay on the colour gently and with care and, if not dark enough, add another layer on top of the first. Smooth the colour to a more even tint by rubbing the surface with a tissue or cotton swab, or by using a 'pastel stump' (from an art shop). Some coloured pencils react better if the swab is first slightly dampened with water, petrol or cigarette lighter fluid, but test them first. Coloured pencils will give patterned effects if a textured surface, such as coarse sandpaper or a textured book cover, is placed beneath the map when colouring. Different textures of the same colour can be used to distinguish related formations and so extend a limited range of colours. Alternatively coloured dots can be added to a lighter base colour with a felt-tipped pen. Dots can even cross geological boundaries to indicate, for instance, a thermal aureole.

Choose colours with care and with due consideration for tradition. In general, use pale colours for formations that cover wide areas and strong colours for those with limited outcrops, such as thin beds and narrow dykes. Always keep your reader in mind: try to follow a system that does not force the reader to keep looking back to the explanation to find out what things mean. Relate your colours to mineralogy, if possible. For instance, if a hornblende schist is shown in pale green, or overlaid with green dots, and a biotite schist is shown in brown, or is overlaid with brown dots, then your readers will probably find your map easier to follow, than if those areas were shown in purple or blue. Note, however, that the use of colour does not obviate the need for formation letters: they also help you to colour your map correctly.

9.8 Stratigraphic Column

In the margin of the fair copy map, a stratigraphic column should be shown. This is the key to the stratigraphic units shown on the fair copy map and on the final cross-sections. It shows all the formations mapped, their map colours, their order (with older units at the base of the column), their full formation names and their thicknesses. The latter is best represented directly in the column by varying the thickness of coloured 'box' for each formation according to some convenient scale (Figure 9.3).

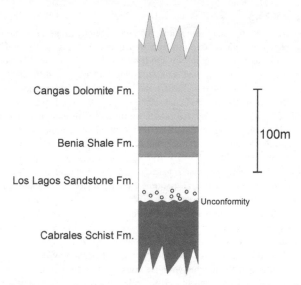

Figure 9.3 Stratigraphic column. Sometimes brief descriptions of the rock-types, fossil content and ages are given on the right of the column.

9.9 Overlays

Do not overcrowd a fair copy map with specialised information in addition to geology, such as rose diagrams, joint measurements and structural statistics. This information is better drawn on transparent paper or film as an overlay to the map. You need not limit the number of overlays. In addition to those mentioned above, they can include: sub-surface contours on specific beds determined from drill holes; isopachytes, isopleths, geochemical contours and even geophysical information. Not only can overlays be superimposed on the fair copy map, they may be usefully overlaid on each other.

An overlay should be the same size as the fair copy sheet and have the same general format. Show the margins of the map area and, because the overlay and the map are of different materials, which distort differently with time, draw 'register' marks to fit the grid intersections on the fair copy.

Title overlays and give them simplified scale bars, a north symbol and an explanation of symbols used. Add a subtitle to indicate to which map the overlay refers and the source of any information that does not originate from you. Add your own name as the author of the map and overlay, and the start and finishing dates of the fieldwork.

9.10 Computer Drafting of the Fair Copy Map

If you are IT literate, an option you may choose is to make a fair copy map using computer drawing software. To achieve this you will need to scan your hand-drawn fair copy map at 150 or 300 dpi using a professional scanner capable of scanning your map, which may be A1 or A0 in size, in one scan. This raster version of your map can be simply geo-referenced by noting the map coordinates of the four corners of the map. Alternatively, you can scan in individual field slips and mosaic them together. Using all the drawing tools available you can trace over the scan to produce a final map with all its boundary lines, symbols, coloured formations plus any map decorations and annotations.

The advantage of such an approach is that the result is of publishable quality in terms of lettering, line work and colours. In addition, the map can be easily modified and redrawn to accommodate new interpretations, new field data, and so on. Multiple copies of your map can also be easily printed. Drawing packages such as CorelDRAW, Adobe Illustrator and AutoCAD allow field slips to be converted into professional-looking fair copy maps.

To make a true GIS version of your geology map is very time consuming. Your hand-drawn fair copy map will have to be geo-referenced and digitally converted into vector data. Every outcrop pattern of each geological unit will be represented by individual geo-referenced coloured polygons with lines representing faults and fold axes, and so on. For a large map this is best achieved by securing your hand-drawn map onto an A0 size digitising table and carefully

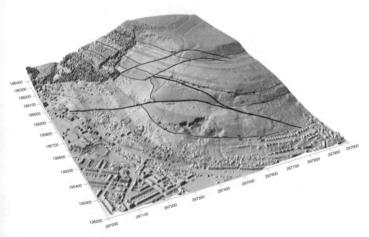

Figure 9.4 *Geological map draped onto a 3D LIDAR digital terrain model, Mynydd Yr Eglwys, Rhondda Fawr Valley, South Wales.*

tracing the outlines of every polygon and lineation using a cursor. Once completed all these geo-referenced polygons and lines can be stored as vector files, which can be analysed, plotted and indexed using GIS software packages such as MapInfo and ArcGIS.

The huge advantage of having a geo-referenced digital version of your geological map is that this allows you to drape your map onto a digital terrain model of your study area (Figure 9.4). It is highly satisfying to see all your hard work in producing a geology map revealed in all its 3D glory. You can then view the map from all directions, zooming in and out or even producing video fly-through animations. As with all computer graphics, the accuracy of the data relies entirely on the quality of the original base map.

10

CROSS-SECTIONS AND 3D ILLUSTRATIONS

Geological mapping involves more than just producing a colourful map to show the distribution of formations within a given area. The trained geologist is able to produce an interpretation of not only the surface rocks but also of their sub-surface structure. This skill of being able to extrapolate to depth is clearly of immense value in the exploration and exploitation of mineral resources, engineering projects, hydrogeology and environmental geology. Young students, embarking on their first mapping exercise, will often ask the supervisor 'How many square kilometres do you want me to map?' The appropriate answer should point out that mapping is a 3D exercise and therefore cannot be quantified in terms of area. Geological mapping is a task that involves gaining information about a certain *volume* of rock.

For the reasons above, no geological map can be considered complete until at least one cross-section has been drawn to show the geology at depth. Cross-sections illustrate the structure of a region far more clearly than a planimetric map. They may be drawn as an adjunct to your fair copy map, and simplified as text illustrations in your report. In addition to cross-sections, columnar sections can be drawn to show changes in stratigraphy from place to place, or 'fence' or 'panel' diagrams to show their variations in three dimensions. Refinements in three-dimensional illustrations include block diagrams, which show the structure of the top and the two sides of a solid block of ground, and models to aid interpretation, such as 'egg crates'. Although much of this kind of drawing can now be done on a computer, you should learn the basis of this kind of illustration by drawing them yourself, and you must also bear in mind that you may not have a computer with you in the field.

10.1 Cross-Sections

Cross-sections are either trial sections, drawn to solve structural problems, or drawn to supplement a fair copy map or illustrate a report. They are also drawn to site boreholes in the search for a lost aquifer or ore body, or by engineering geologists planning the excavation of a tunnel. The cross-section shows the interpreted position of formation boundaries at depth, but should be also used to display faults, cleavage and the axial surfaces of large folds. In tectonically

Basic Geological Mapping, Fifth Edition.
Richard J. Lisle, Peter J. Brabham and John W. Barnes.
© 2011 John Wiley & Sons, Ltd. Published 2011 by John Wiley & Sons, Ltd.

interesting areas, it is useful to show internal structures such as bedding within the formations in addition to the formation boundaries.

Cross-sections are often poorly drawn, usually for one or more of the following reasons:

1. The information on which the cross-section is based comes entirely from a small part of the map – a narrow strip of ground along the line of section. Take a broader view; examine the whole map. Structural relationships in the cross-section should reflect those seen on the map. For example, if a given formation varies significantly in thickness across the area, it will also show thickness variations in the cross-section. If the outcrop of a thrust fault tracks along parallel to stratigraphic contacts it probably should do so in section too (see Figure 6.20).

2. The interpreted structure relies too heavily on measured dips of strata (Figure 10.1). Suspect interpretations arise from blindly projecting the dips measured at the surface, to depth. This can lead to formations wedging out in the cross-section, an interpretation that may not agree with evidence from the map (Figure 10.1a). To remedy such problems remember that the dip may change with depth, giving curved contacts in cross-section that still honour the dips measured at the surface (Figure 10.1b).

3. The interpretation does not take full account of outcrop scale observations. Minor structures such as the bedding–cleavage relationship and the asymmetry of minor parasitic folds provide invaluable clues to the major structure and this information should be used in your interpretation of the cross-section (see Section 6.5).

Furthermore, the style of folding and faulting seen at the outcrop is the best available indicator of the structural style on a larger scale. For example, if thin competent beds show folds of parallel style on a small scale, it is more than likely that competent formations parallel folds of parallel type on the scale of the section (see Section 6.5.5).

10.1.1 Trial cross-sections

Draw a cross-section whenever a problem of interpretation arises. Do it whilst still in your field camp so that you can make the additional structural measurements if needed (see Section 8.2.5). Even when you have no problems, solutions should be drawn during the fieldwork stage to ensure that nothing is going to be missed. In geologically complex areas there may be more than one interpretation of the structure, and trial cross-sections will at least show which is the most probable. Drawing cross-sections should become second nature to a geologist.

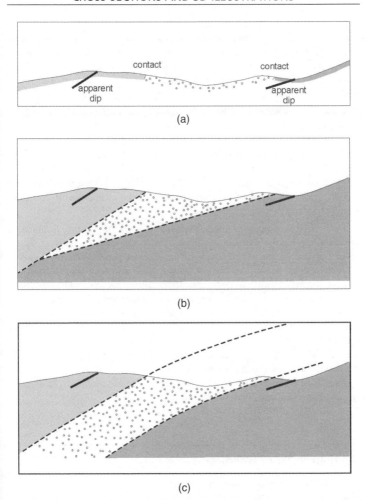

Figure 10.1 *Structural interpretations. (a) Data consisting of contact positions and apparent dips. (b) Interpretation assuming maintenance of dips at depth. (c) Interpretation assuming constant thickness of units. Relationships on the geological map will help decide between these alternatives.*

10.1.2   Fair copy cross-sections

One or more fair copy cross-sections are drawn to accompany a fair copy map. The number of cross-sections required depends on the complexity of the geometry of the structures. Draw them to the same standards and colour it with the same tints as your map, for they are to all intents and purposes part of that map; you may even redraw them later in the margin of your final map.

Display cross-sections in the map margins wherever possible, so that all geological information is kept together. If sections have to be drawn separately, draw them all on the same sheet so that they can be easily compared. Show on the fair copy map the positions of all sections presented by lines drawn on the face of the map, with the ends of each section line clearly indicated by a short cross-line, as in Figure 9.1. Geologists should curb their imagination on fair copy sections and not show interpretations down to improbable depths for which they have no possible evidence.

Draw cross-sections as if you are looking in a general westerly or northerly direction, so that the southern, south-western and western ends of the section always appear on the left-hand side of the sheet of paper. Whenever possible draw them to go across the strike of beds as close to a right angle as possible; if there is a broad swing in strike across the map, change the direction of your section line at one or two well-separated points to keep it as nearly perpendicular to the strike as possible. The scales vertically and horizontally should be stated. Normally, to avoid problems of distortion of the geological structures in the section, horizontal and vertical scales should be the same. Exceptionally where dips are no more than $10°$, an exaggerated vertical scale may be permissible, but always state the true dip on the section.

Simplified cross-sections are also frequently used as text figures to illustrate specific structures described in a report.

A checklist for cross-sections

1. Is the line of section marked on the map?
2. Is the line of section well chosen for displaying the structure of the area? In many cases more than one section is required.
3. Does the line of section have kinks in it? If these changes of direction are too large, the resulting structure seen in the cross-section may be misleading.
4. Does the variation of formation thicknesses look realistic?
5. Do the relationships of the units on the map appear in the cross-section?
6. Are the vertical and horizontal scales the same? Are they stated on the cross section?

174

7. Are the scales of the map and the cross-sections the same?
8. Does the structural style in the section agree with that of smaller structures seen in the field?

10.1.3 Serial cross-sections

Serial cross-sections are drawn along regularly spaced parallel lines, usually on large-scale plans used for mining or engineering purposes. They may be drawn at right angles to the strike of the structure but more usually they are drawn parallel to one set of grid coordinates. Their object is to show progressive changes in the geometry across the area (see below, Section 10.2.3 and Figure 10.5).

10.2 Method of Apparent Dips

Poorly drawn cross-sections are so often encountered in professional life that a resume of the process is given below, although most readers will have already been taught this method early in their geology courses.

1. Draw the line of section (A–A′) on the face of the map, marking each end of the line with a short cross-line (Figures 9.1 and 10.2).
2. Fasten the map to a drawing board or table with the section line parallel to the bottom edge of the board/table,
3. Tape to the map, a few centimetres below the section line, a strip of paper on which to plot the section.
4. Draw a base line on this strip of paper parallel to the section line on the map. Then draw a series of parallel lines at the chosen contour interval above it. The spacing of these lines has to accord with the vertical scale (Figure 10.2a).
5. Tape down a plastic ruler or steel straight-edge so that it cannot move, well below and parallel to the base line.
6. By sliding a set-square (triangle) along the straight-edge, drop a perpendicular down to the appropriate elevation on the section paper from every point where the section cuts a contour line on the map (Figure 10.1a). Join these points to give the profile of the topography.
7. Extend the strike line of any strike/dip symbol until it meets the section line on the map, for any strike symbol lying close to the cross-section (Figure 10.2b). The distance you may project a strike is a matter of geological judgement. Where there is obvious flexure, extend the strike line to follow its curve to meet the line of section.
8. Drop down this point of projection on the section line onto the topographic profile using the method of step 6 above, and then plot the apparent dip in the line of section (see Section 10.6.2).

175

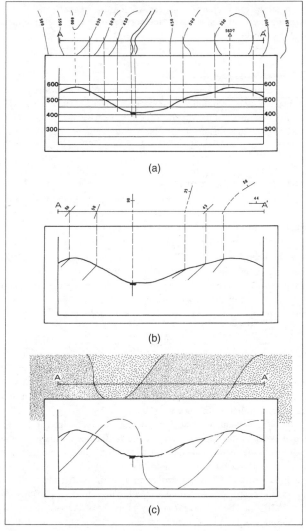

Figure 10.2 *(a–c) Drawing a cross-section from apparent dip information. See text for details.*

9. Still using the set-square, drop a perpendicular wherever the section line crosses a geological contact on the map and lightly mark the position on the profile of the topography.

10. Lightly sketch in the structure by extending 'dip lines' and drawing contacts parallel to them. Then modify the interpretation to allow for thickening or thinning of beds, and for any further suspected change in straightforward folding or tilting. Do not interpret geology to improbable depths beneath the surface. Complete your interpretation by continuing the structure above the topographic surface; you have just as much evidence there as for your interpretation below the surface (Figure 10.2c). Finally, ink in your interpretation including, where appropriate, dashed lines for parts of the structure now eroded away above the land surface.

The task is made easier if a 'T' square or drafting machine is available.

There are other methods of constructing cross-sections appropriate for use in different types of structural settings, including geometric constructions (Badgley, 1959; Marshak and Mitra, 1988; Rowland et al., 2007; Ragan, 2009) such as the Busk Method and the down-plunge projection method (Section 10.3).

10.2.1 Calculating apparent dips

Unless a cross-section line cuts the strike at right angles, the angle of dip must be modified in the cross-section because apparent dip must always be less than true dip; this is purely a matter of geometry. Apparent dip can be determined by graphical methods, trigonometric methods (Lisle, 2004; Ragan, 2009), or by conversion tables or charts as found in most books on structural geology (e.g. Berkman, 2001). The stereonet, the geologist's geometrical calculator, can also be used (Lisle and Leyshon, 2004).

10.3 Down-Plunge Projection Method

The down-plunge method of cross-section construction is the preferred method in areas where the geological structure is dominated by folding and where the folds have hinge lines with a constant plunge (in angle and trend). The method assumes that the geometry of the structures remain unchanged if followed in the direction of the fold hinge lines. This allows points lying on a geological contact on the map to be projected in the direction of the hinge lines onto the section plane. The construction steps are:

1. Draw the selected line of section A–A' on the map, a cross-section template with a baseline of length A–A' and a set of horizontals representing different elevations (Figure 10.3). The spacing of these horizontals is determined by the map scale.

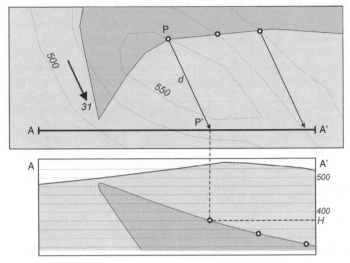

Figure 10.3 *Drawing a cross-section using the down-plunge method. See text for details.*

2. Select a geological contact on the map.
3. Choose an arbitrary point P on the contact and draw a line through it in the plunge direction to meet the line of section at P'. P' is the projected position of P in the cross-section.
4. Measure the distance d from P to P' in real units. The distance d is positive or negative depending on whether the plunge is from P towards P' or from P' towards P, respectively.
5. Plot the projected point in the cross-section at an elevation H given by:

$$H = h - d \tan(\text{angle of plunge})$$

where h is the elevation of the ground at P (rather than P').
6. Repeat steps 3–5 for other points on the contact, and join the projected points in the cross-section. This shows the folded shape of the contact in the cross-section.
7. Repeat for other boundaries.

An important advantage of this method is that the cross-section is constructed from information taken from across the map, not just from along the line of section. The method relies on knowing the plunge and plunge direction of the folding. This information can be obtained by direct measurement of minor fold hinges, or by stereographic analysis of strike and dip readings (see Lisle and Leyshon, 2004, p. 44, for details).

10.4   Balanced Cross-Sections

A balanced cross-section is one constructed using assumptions relating to the compatibility of structure interpreted at different stratigraphic levels. For example, one method of constructing balanced cross-sections assumes that the lengths of different formation boundaries drawn in the cross-section should be equal, if allowance is made for folding and faulting. This assumption constrains the range of possible structures that also fit the data from dips, outcrop positions, and so on. Details of these methods can be found in McClay (2003).

10.5   Columnar Sections

Columnar sections consist of a number of simplified stratigraphic columns shown side by side to illustrate how stratigraphy changes from place to place (Figure 10.4). They are prepared from surface outcrops and drill-hole logs.

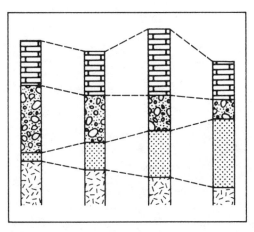

Figure 10.4 *Serial cross-sections, in which each columnar section is constructed by oblique projection.*

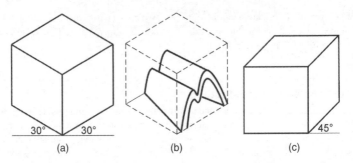

Figure 10.5 *(a) Isometric cube; (b) a fold drawn using the isometric cube as a guide and (c) oblique projection of a cube.*

10.6 Block Diagrams

Three-dimensional diagrams greatly help readers of reports to visualise the subsurface geology of the area concerned. Their preparation may help your own understanding too. There are two basic methods for projecting features onto the visible faces of the block: *isometric projection* and *oblique projection*. Both are simple to construct.

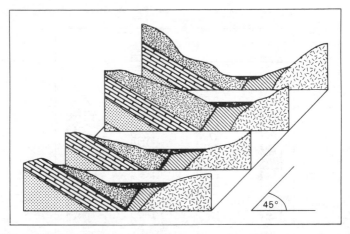

Figure 10.6 *Oblique projection of serial cross-sections. Each cross-section is a true (undistorted) cross-section but the distance between them is foreshortened by one-third.*

180

10.6.1 Isometric projection

Unlike true perspective, isometric projections have no vanishing points: all parallel lines remain parallel in the diagram. The two horizontal coordinates are inclined at 30° as shown in Figure 10.5. The viewer sees the faces of a cube, for instance, as three equal-sided parallelograms (Figure 10.5a). Geological information drawn on the faces of an isometric block must be distorted to fit them. Do this by drawing a grid over your geological information, map or cross-section, and also on the faces of the isometric block, then transfer information on the block by eye, grid square by grid square or by rectangular coordinates. Isometric paper can be bought and is a useful aid to drawing such diagrams.

You can also use isometric projection as a framework for 3D illustrations, showing the rock structure within a 'see-through' cube. Figure 10.5b shows the principle.

Using the same method, block diagrams can be drawn that represent rectangular blocks instead of cubes. They can also be split into two or more slices, or

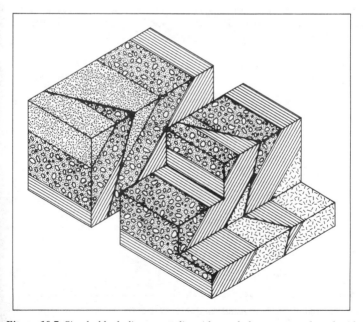

Figure 10.7 Simple block diagram, split, with two halves separated, and with steps.

have steps cut into the block to illustrate specific points, but avoid overhangs (Figure 10.6).

In the isometric block diagram the imaginary cube of rock is viewed in the direction of its diagonal. Block diagrams with other directions of viewing can also be constructed (Lisle, 1980; Ragan, 2009).

10.6.2 Oblique projection

In oblique projection, the front face of the basic cube is an undistorted cross-section in the plane of the paper (Figure 10.5c). The side face of the cube, however, is a parallelogram inclined at 45°, with distances receding from the viewer foreshortened by one-third to prevent the side appearing rectangular. Oblique projection can be used to draw a block diagram; another use is to draw serial cross-sections (Figure 10.7).

10.6.3 Fence (panel) diagrams

Fence diagrams, as distinct from columnar sections, are three-dimensional illustrations. Stratigraphy or lithology is shown on 'fences', or 'panels', connecting the different sites. Again, either isometric or oblique projection can be used (Figure 10.8). Their principal use is in plotting information from borehole logs.

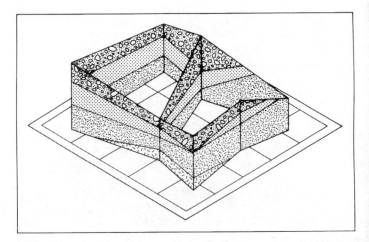

Figure 10.8 *Fence, or panel, diagram illustrating stratigraphic variations in a region. It gives a better spatial impression than a columnar section (cf. Figure 10.4).*

10.7 Models

Geological problems can sometimes be solved by constructing physical three-dimensional models that can be viewed from every direction. They need no embellishment if used only as interpretative aids.

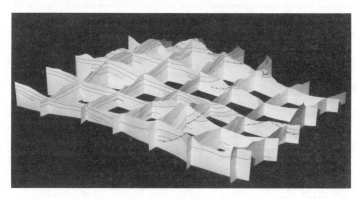

Figure 10.9 *Photograph of an egg-crate model (by courtesy of S.J. Matthews).*

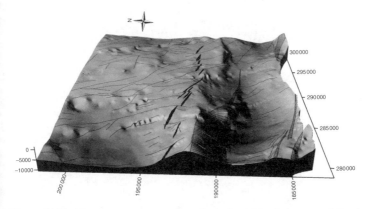

Figure 10.10 *3D computer-generated geological model of the central South Wales coalfield created from subsurface geological information (from Brabham, 2009).*

10.7.1   Egg-crate models

An egg-crate model is constructed from a series of intersecting cross-sections. Draw the sections on heavy cartridge paper or Bristol board, then cut them out and slot them together by cutting slits where they intersect. Such models are particularly useful in mountainous areas where the geology has been complicated by nappes and overthrusting (Figure 10.9).

10.7.2   3D block modelling

Visualising and understanding a complex 3D subsurface structure can be very difficult, especially if you have to give a presentation to a non-technical audience. To achieve this, geologists have tried to create block models to help visualise the 3D structure. In the past this was achieved using physical models made from clear Perspex. Geological sections were drawn onto individual Perspex sheets. These vertical and horizontal geological sections were then locked together to create a physical model. In 3D mining models, pipes were often run through the model to represent the labyrinth of underground mine workings.

Such physical models have now largely been replaced by 3D computer visualisations using specialised software such as RockWorks, Maptek Vulcan, Surpac and Datamine. Getting all the available geological data into a 3D computer format can be very time consuming. However, once the computer model has been produced, it can be rotated and viewed from any angle. Video animation 'fly-throughs' can also be created by the software for presentations. Using 3D laser cutting technology, plastic physical models can also be created from the digital data.

11

GEOLOGICAL REPORTS

The completion of any period of geological mapping is usually followed by the production of a geological report to explain the geology of the region investigated. In fact, professional geologists probably spend more time preparing reports (with the attendant laboratory work and map drafting) than they do in the field.

Regarding the appropriate style to be used for your report it should be remembered that your mapping project is a scientific investigation and its findings should be laid out in the manner of a scientific publication. Essentially this involves first explaining what was done, then stating what observations were made and finally what conclusions were reached.

The object of writing a report, or for that matter a paper or thesis, is to communicate ideas to others. Although there is no reason whatsoever why geological writing should not be in good, clear English, it seldom is. 'Unclear English and the inability to be brief' are major complaints among employers and clients (Opinion, 2002). Geological literature is all too often obscure, ambiguous, circumlocutory and irritating to read. Sad to say that whilst most young geologists take trouble to improve their mapping techniques, they appear reluctant to improve their ability to produce readable reports. This brief chapter is but an introduction to report writing; fuller information can be found in the many excellent books on the subject. First, however, every geologist, young and old, should read Vansberg's 'How to write geologese' (1952), a beautifully researched tongue-in-cheek letter to *Economic Geology* in which he quotes many extracts from the geological literature. In it we can all recognise many of our own mistakes in examples from other writers. Among books that can be recommended are Hansen (1991) and Barrass (1978). In addition, the Penguin edition of *The Complete Plain Words* (Gowers, 1986) originally written to teach civil servants to be comprehensible, is an amusing guide to sensible grammatical English. There are many more recent books on the same subject, mostly addressed to scientists and engineers, who all suffer from incomprehensibility.

Basic Geological Mapping, Fifth Edition.
Richard J. Lisle, Peter J. Brabham and John W. Barnes.
© 2011 John Wiley & Sons, Ltd. Published 2011 by John Wiley & Sons, Ltd.

11.1 Preparation

It is just not possible to sit down and write a report at one sitting. If you do try, your report will be incomplete and badly written; report writing takes effort. First you must plan the layout, section by section, then draft each section using all your notes, maps, laboratory results and references gathered from other sources. This is the stage where you get your thoughts down 'on paper' without too much regard to grammar and literacy; there is not even any need to draft the sections in the order they will appear in the finished report. Sections can be rearranged and added to easily with the word processor. As you write, list the illustrations needed to support the text; they can be roughed out whenever you need a break from writing.

11.2 Revising and Editing

Your first rough draft gets the essential facts and information in order and allows your ideas to develop. The next stage is to revise this first draft. Earlier ideas may need to be changed, new ideas may emerge; but, still, more attention should be given to ideas than to the English. The next stage is to edit the draft. The order of some paragraphs may need changing, spelling mistakes corrected and grammar improved. Indicate where illustrations should be inserted. Now read it through as a whole and improve the English: this depends more on hard work than art; refer to dictionaries and check references, you may need more than one revision to get the right result. Ernest Hemingway rewrote the last page of one book 39 times before he was satisfied. Make a final check of the whole report for punctuation, and ruthlessly prune anything that does not contribute to meaning. Aim at concise, direct, plain English. Look especially for repetition; geologists have a blind spot for some words that they use over and over again, with irritating monotony. 'Area' is one of them, for example '...the Falmouth *area* was part of the *area* mapped and covered 10 km² in *area*'. Such a sentence is not unusual: it needs recasting. The last 'area' in the sentence is redundant, and 'region' or 'district' could be substituted for at least one of the others. Most repetitions can be eliminated by rewriting, but not all repetition is bad. Occasionally it can be used for emphasis and it is always better than ambiguity.

11.3 Layout

Any scientific report can be broken down into a number of basic sections or chapters:

Title page
(Contents)
Abstract

Introduction
Main body
Conclusions
References
Appendices

In short reports, some of these parts may be only a few lines long. In others, even the *Introduction* may include several sub-sections, and the *Main body* itself may consist of many sections and sub-sections. The composition of each part is described below.

11.3.1   Title page

This is more important than many writers think. It shows what the report is about, who wrote it and when. Lay it out carefully, for it is the first thing your reader will see, and first impressions are always important. Packaging it may be, but if you do not think your report is worth presenting properly, why should your reader?

11.3.2   List of contents

Any report longer than 20 pages needs a list of contents to show readers how the subject has been covered and to help them to locate information they may wish to refer to again.

11.3.3   Abstract

The *Abstract* (or *Summary*) follows immediately after the title page (and Contents list, if any). It must be written last, after you have already formulated your ideas and conclusions. This does not mean that it can be dashed off as an afterthought. It must be as literate as the report itself and concisely review the work done and its results. Preferably, it should not exceed 200 words and, in a short report, very much fewer.

11.3.4   Headings and sub-headings

The body of the report will usually be divided into sections, and further subdivided into two lower categories, each with its own type of heading; the main sections are equivalent to the chapters of a book. Each section starts on a fresh page with its headings capitalised and centred on the page. Sections contain sub-sections with their headings either centralised or at the left-hand margin, depending on the house rules of your employer, written using upper and lower case. Do not underline headings, because if your report (or paper) is sent to a printer for publication, underlining tells compositors to print those words in italics. Sub-sub-headings break sub-sections up into smaller units, and these headings are again typed at the left-hand margin of the page and usually italicised, sometimes followed by a dash, colon or a wider space. With the widespread use of computer word-processing software, you can of course produce your own italics, and also bold type, and larger font to emphasise the hierarchy of headings. Here we only make generalised suggestions:

9 Section Heading
9.1 Sub-section Heading
9.1.1 Sub-sub-Heading

By an extension of the system, if further sub-divisions are needed, logically they should be numbered *9.1.1.1 . . .* or even *9.7.11.5*. Such designations are unwieldy and it is better to use lower case Roman numerals (i, ii, iii, etc.) or bracketed lower case letters – i.e. (a), (b), (c), etc. This category of heading should only be needed occasionally.

Plan your headings from the start as part of an outline for your report. Commercial companies often provide staff with a detailed list of headings and sub-headings to cover every eventuality, often many pages long, to ensure nothing is forgotten, although not every heading is expected to be used in any single report.

11.4 The 'Introduction'

A report needs an Introduction so that your readers know what it is about. They need to know what you did, why you did it, when you did it and where you did it. They also want to know what has been done before and by whom. Include a small index map to show where the area is in relation to the region around it, and its general geography, topography and communications, how to get there and the main place names. The Introduction usually includes a summary of the geological context, without going into detail. It sometimes also gives a brief review of the vegetation, land use and economy of the region, emphasising aspects that are geologically related. The Introduction is also a useful place to acknowledge any help given to the writer, both in the field and in the preparation of the report.

11.5 Main Body of the Report

Hard and fast rules cannot be given on what should be included in a report – that depends on the subject; but many reports are basically a description and explanation of the geology of a limited area, sometimes covering only a few tens of square kilometres. Normally, the main body of such a report will consist of sections with headings similar to those shown below.

11.5.1   Regional geology

Before embarking on a detailed account of the geology mapped, it helps the reader if the main features of the area are outlined first. In very short reports, general geology can be included in the Introduction, together with the regional geology. In longer reports, a separate section is needed, supported by at least one text figure showing the base outlines of the main geological units on a small-scale map, preferably with some indication of structure and place names. Note that the US Geological Survey insists that any place name mentioned in the text must appear on some map within the report.

11.5.2   Stratigraphy, and so on

This is a substantial section dealing with a systematic description of the mapped formations. In an introductory section of this chapter, a text figure showing the geological succession of formations and their thicknesses, coloured with the same tints as those on the main geological map, is a great help to the reader. This is followed by sub-sections each devoted to a separate formation. If the relative ages of the formations have been determined this should govern the order of these sub-sections (older first). If ages are not known, the structural order is used with the lowest units first. The rocks may be described in much the same way as they are described in your field notes (see Section 7.2) except now you also have the benefit of examining thin sections, laboratory work, consultations with colleagues and references to literature. A description is required of the formation's:

- distribution across the area;
- lithological characteristics, and their variation laterally across the area and from base to top of the formation;
- formation thickness and its variation across the area;
- defining features of the contacts between formations;
- topographic expression;
- sedimentary structures;
- fossil content.

These descriptions should be followed by interpretations based on your observations, for example age, environment of deposition. Here is an example of a description of a fictitious formation:

The Tarna Conglomerate Formation

This is a thick clastic unit that conformably overlies the Puebla Shale Formation. Its outcrop occupies an area of some 20 square kilometres in the north and west of the mapped area. Thickness is estimated to be about

600 m. The type section is just north of the village of Altasil. The lower 150 m is composed predominantly of grey and buff argillaceous quartzite. Some quartzite beds are conglomeratic. Clast-supported grey conglomerates as thick as 6 m are present. Clasts are dominantly quartzite, but about 3% are composed of grey limestone. The lower unit is overlain by a 450 m thick clast-supported conglomerate, with predominantly limestone clasts and in a poorly sorted matrix.

The deposition of the Tarna Conglomerate Formation represents a marked change from the low energy marine of the Puebla Shale Fm and could be fluvial and/or shallow marine in origin.

11.5.3 Structure

Regional structure has already been introduced under the earlier report section 'Regional Geology'. Now you describe the more specific structural details of the area mapped, based on your own field evidence. Usually the following are described:

1. The minor geological structures. As usual, describe the structural features, for example bedding, cleavages, faults, joints and so on, that were directly observed in the field, before mentioning broader conclusions. Mention the appearance of the structures, their scale, the development in relation to rock-type and their orientations. Stereograms of the measured structures are useful for the latter (Lisle and Leyshon, 2004). Figures consisting of field sketches and photographs will assist with these descriptions.
2. The evidence of age relationships between the different structures and with the sedimentation.
3. The evidence for any major structures deduced from map patterns and analysis of minor structures, for example parasitic folding, cleavage. A structural summary map included as a text figure is useful. Major structures could be named on such a map, for example the Vargas Thrust, Arenas Anticline and so on, because this would make it easier to refer to these structures later in the report. Summary cross-sections presented as text figures will also simplify this description. Structural geology is an excellent example of a subject where a diagram can save a wealth of text. Isometric diagrams are especially helpful and are much easier to draw than they look, and even more so using a computer (see Section 10.7.2).

11.5.4 Metamorphism

Metamorphism may deserve a section of its own, but logically, it may often be part of the structural section. The way in which these two subjects are treated is a matter for the judgement of the writer.

11.5.5 Igneous activity

Igneous activity covers a wide spectrum, from plutonism to vulcanicity. Intrusive bodies can be dealt with separately from the main stratigraphy, but using the same descriptive criteria mentioned above. Be sure to discuss the cross-cutting relationships, which give you evidence of the age of the intrusion.

11.5.6 Economic geology

All too often economic geology is glossed over. Quarries will normally be examined during fieldwork, if only because rocks can be seen in them. Sand, gravel and clay pits are, however, seldom mentioned in reports, yet they are mineral assets of the region. If metallic minerals occur, they relate to the geology. The potential value of other mineral materials must not be ignored either. Limestone, for instance, is often described in considerable academic detail in reports, yet its suitability as a cement material, flux, industrial chemical, pigment, decorative stone or it many other industrial uses seldom gets a mention. Refer to Scott and Bristow (2002) for the uses and geology of industrial minerals. Water is another natural resource all too often ignored.

11.6 The 'Conclusions' Section

The foregoing sections of the report are mainly factual and depend on observation, supported by interpretation based on established geological processes. Major hypotheses may have been avoided so far. Now, however, results are brought together and conclusions drawn from them. Controversial ideas should be explained with mention of evidence in support and against. You may have alternative explanations for your observations. These, and their relative merits, should be described.

Sometimes this section is concerned for a great part with the geological history, but it need not deal exclusively with this. In more specialised reports, the conclusions may be of a different character, and may also include recommendations to tell the reader what she/he should do next, why, how it should be done and sometimes how much it will cost. In professional reports the recommendations may warrant a section of their own.

11.7 Text Illustrations

Geological reports are greatly improved by the inclusion of illustrations. The latter are all referred to as *figures* even though they take a variety of forms:

- Line drawing based on a labelled field sketch (redrawn from your notebook).
- Line drawing to illustrate a proposed concept, for example the evolutionary stages of development of a geological structure.

- Large-scale map showing geological detail of a small area (e.g. Figure 4.16).
- Cross-section of some notable structure.
- Sedimentary log, graph, histogram, rose-diagram or stereogram.
- Photograph or explanatory sketch of photograph.

The preferred style of field sketches is discussed in Section 8.2. Conceptual drawings should be kept simple with labelling for clarity. Where necessary, include geographical coordinates and scale. If you have access to computer drawing software such as CorelDRAW or Adobe Illustrator, then publication quality figures can be produced and pasted into the document file containing the text of the report.

Photographs, and all other figures for that matter, are not added to the report for general embellishment. They should only be included in order to illustrate a specific point. All figures therefore have to be referred to in the text, and are numbered according to the order in which they are mentioned.

Figures are referred to within the text in two ways:

Figure 3.11 shows a typical exposure of the Benia Limestone Formation with alternations of thinly-bedded limestone and shales...
Cross-bedding within the quartzite is common (Figure 3.16) and provides evidence of way-up.

All figures should have a caption that explains the importance of the figure.

11.8 References

Any reference you make in the text to previous work done by anyone must be acknowledged, whether the information is from published work, an unpublished report or merely by word of mouth. This is a matter of scientific ethics. There are accepted forms of referencing. The Harvard system is one used by the Geological Society of London and is suitable both for manuscript reports and for publications. Briefly, a reference is acknowledged in the text by the name of the author and the year in which it was published.

11.8.1 Referring to publications in the text

References in the text to papers with a *single* author should be cited as 'the work of Baker (1971) shows that...' or 'recent surveys (Baker, 1971) show that...'. References in the text to papers with two authors should be cited as 'Glover & Robertson (1998) show that...' or 'these results are supported by other workers (Glover & Robertson, 1998)'.

References in the text to papers with *more than* two authors should be made thus: '...(McIlroy *et al.*, 1998)...', but cited in full in the bibliography.

11.8.2   Reference list (bibliography)

References are listed at the end of the report in alphabetical order of the first author's surname of each entry, and where there are multiple authors, all are listed; *et al.* is not used in the references list. The following fictitious examples cover most types of publication to be found in a reference list, and should be cited in the bibliography as below:

Baker, J. W. 1971, The Proterozoic history of southern Britain. *Proceedings of the Geologist's Association*, 82, 249–266.

McIlroy, D., Brasier, M. D. & Moseley, J. M. 1998, The Proterozoic-Cambrian transition within the 'Charnian Supergroup' of central England and the antiquity of the Ediacara fauna. *Journal of the Geological Society of London*, 155, 401–412.

If you are referring to a book, this is the standard format:

Press, F. & Siever, R., 1986, *Earth*; 4th Edition., Freeman and Co., New York.

If you are referring to a book with separate authors for each chapter, edited by another person(s), this is the standard format:

Sparks, S. J., (1992), Chapter 5: Magma generation in the Earth. *In*: Brown, G. C., Hawkesworth, C. J. & Wilson, R. C. L. (eds) *Understanding the Earth: A New Synthesis*. Cambridge University Press, 91–114.

11.9   Appendices

Appendices, or appendixes (both are correct), contain the 'unreadable' factual evidence on which many reports rely but which is difficult to include in the text. It includes long lists of analytical data, statistical information, sample localities, and graphs and curves used in standardising instruments. It may even include commissions and other letters. Some industrial reports consist of more appendices than text. Appendices must be properly arranged and any explanatory matter should be just as literate as any other part of the report. Appendices are not junk heaps for the haphazard collection of inconvenient material.

11.10   Some Final Thoughts

This brief review is intended only as an indication of how a report should be written. Report writing is such an important adjunct to geology fieldwork that students should be actively encouraged to study the subject in just as much detail as any other branch of the science. Unfortunately, this is seldom done and graduates are usually left to learn for themselves after graduation, much to the exasperation of their employers. Finally, those about to write a report might well take heed of what Samuel Coleridge said 180 years ago: 'if men would only say what they have to say in plain terms, how much more eloquent they would be' (Barrass, 1978). In other words, write in direct plain English; avoid journalistic, civil service and geological jargon: prune out unnecessary words and phrases and write for your readers' benefit not your own.

Appendix A

ADJUSTMENT OF A CLOSED COMPASS TRAVERSE

A compass or levelling traverse will seldom close without error. For example (Figure A.1), a traverse starts at point A (black triangle), passing through intermediate points B, C, D, E to the final point F. The survey is closed by surveying from F back to the start at A. The survey is then closed by surveying from F back to A. In theory the location of A should plot to the same point; however, when you plot these points on a map there will be a *closure error* and A will plot on the map at A′, so the horizontal error is A−A′.

Let's assume this error works out as 110 m. You can adjust this error back through the survey by drawing lines parallel to A−A′ through each of the intermediate points, then distribute the error of 110 m at each intermediate point in

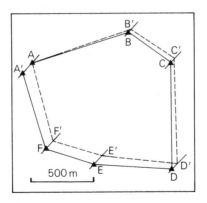

Figure A.1 Adjustment of a closed compass traverse.

Basic Geological Mapping, Fifth Edition.
Richard J. Lisle, Peter J. Brabham and John W. Barnes.
© 2011 John Wiley & Sons, Ltd. Published 2011 by John Wiley & Sons, Ltd.

proportion to the *total* distance travelled to reach that point (plotting B′, C′, D′, E′ and F′).

Closure error A–A′ = 110 m

Total distance travelled = 3000 m

Correction factor = 110/3000 = 0.03 m/traverse

Appendix B

FIELD EQUIPMENT CHECKLIST

Forgetting to pack essential items of equipment can cause inconvenience, and sometimes a wasteful loss of field time. This comprehensive checklist should be consulted before departure, though not every item is required for every period of fieldwork.

Mapping equipment

Safety helmet
Protective goggles
Rucksack
Map case
Notebooks
Pencils for plotting
Knife/sharpener
Coloured pencils
Technical pens
Scales
Protractors (half-round; 15 cm diameter and 10 cm spare)
Pencil case (for belt, chest pocket or map case)
Hammers
Chisels
Belt and hammer frog
Pocket tape
Tape, 30 m
String or cord
Field acid bottle
Compass/clinometer/hand-level
Camera, tripod
Filters (esp. UV)
Memory cards, flash equipment
Binoculars
Pocket stereoscope

Basic Geological Mapping, Fifth Edition.
Richard J. Lisle, Peter J. Brabham and John W. Barnes.
© 2011 John Wiley & Sons, Ltd. Published 2011 by John Wiley & Sons, Ltd.

Hand-lens plus spare
GPS instrument
Specimen bags
Newspaper for wrapping specimens
Boxes/cotton wool for fossils
Felt-tipped pens
Charts, tables, stereonet.

Sampling equipment

Entrenching tool
Trowel
Shovel/pick
Chisels/moils
Auger
Sieves
Gold pan
Camel-hair brush
Tubes for concentrates
Funnel

Rucksack kit

Spare sweater and socks
Waterproof anorak/cagoule
Waterproof trousers
Leggings
Lunch box
Vacuum flask
Water bottle
Tin/bottle openers
Corkscrew (France?)
Knife (Swiss army?)
Insect repellent
Suncream
Lip salve
Toilet paper

Rucksack emergency kit

Whistle
Torch/batteries
Small spare compass
Radio beacon/flares

Mobile phone
Space blanket
Emergency bivouac ('polybag')
Emergency rations (chocolate, etc.)
Waterproof/windproof matches
Water purifying tablets
First-aid kit (plasters, etc.)
Army-type sealed field dressing
Sealed antiseptic wipes

Field clothing (temperate and cold climates)

Anorak/cagoule/padded jacket
Sweaters
Socks (ample supply)
Boots
Rubber wellington boots
Gloves and spares
Shirts (long- and short-sleeved)
Trousers/jeans/chinos
Woolly hat
'Long johns'/vests, and so on
Mosquito face net (Arctic areas)

Field clothing (warm climates)

Shirts (long- and short-sleeved)
Jeans/chinos/shorts
Jungle hat
Sunglasses

Drawing, plotting, 'office' equipment

Maps (road, district, etc.)
Maps for plotting on
Aerial photographs
Handbooks (geology)
Reference manuals
Permatrace, Mylar, tracing film, tracing paper
Squared paper
Stereonets
Probability paper
Pocket calculator
Drafting tape

Waterproof ink
Mapping pens (for very fine work)
Stylus-type pens (black/colours)
Straight-edge
60/30 set-squares
45/45 set-squares
Map scales
Craft knives/erasing lances (razor blades?)
Pencils: 2H, 4H
Erasers
Emery boards for pencil sharpening
Coloured pencils
Pastel stubs for smoothing colours
Needle for pricking through aerial photos
Scissors
Mirror stereoscope
Writing paper
Pads of paper for drafting reports
Envelopes
Stamps
Laptop computer

Items for camp use

Tent and pegs
Bed/bedding
Mosquito/sandfly net
Fly spray
Insect-repellent cream
Primus stove
Pressure lamp
Paraffin/kerosene
Candles
Water container
Water filter
Cooking utensils
Plates, cutlery, and so on
Tables, chairs
Washbasin/camp bath
Toilet soap, towels
Soap for clothes washing
Toilet paper

Pick and shovel/entrenching tool (for digging pits for latrines
    and rubbish disposal)
More comprehensive first-aid kit
First-aid manual

Paperwork

Passport, with at least six months to run
Visas
Vaccination/inoculation certificates
Driver's licence
International driving permit
'Green card' insurance
Car spares
Tickets
Foreign currency
Traveller's cheques
Cards: debit and credit
Any authorisation, work permits
Phrase book/dictionary

Also!

Start taking antimalarials well ahead of departure.
Leave your address and contact phone numbers so you can be contacted.
Leave dates of going and return.
Pay all bills.

Appendix C

INDICATORS OF STRATIGRAPHICAL WAY-UP

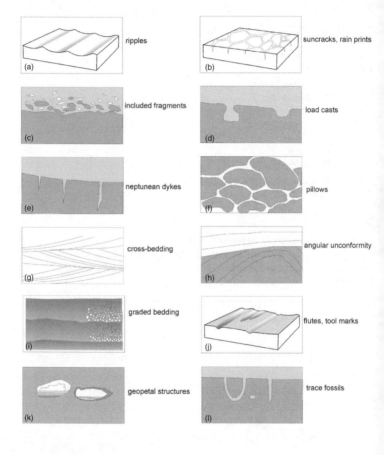

(a) ripples

(b) suncracks, rain prints

(c) included fragments

(d) load casts

(e) neptunean dykes

(f) pillows

(g) cross-bedding

(h) angular unconformity

(i) graded bedding

(j) flutes, tool marks

(k) geopetal structures

(l) trace fossils

Basic Geological Mapping, Fifth Edition.
Richard J. Lisle, Peter J. Brabham and John W. Barnes.
© 2011 John Wiley & Sons, Ltd. Published 2011 by John Wiley & Sons, Ltd.

Appendix D

USEFUL CHART AND TABLES

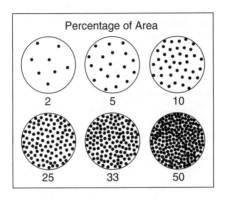

Chart D.1 *Percentage area chart.*

Table D.1 *Spacing of bedding and jointing.*

Spacing (mm)	Bedding	Jointing
0–1	Very thinly laminated	–
1–10	Thinly to thickly laminated	–
10–100	Very thin to thinly laminated	Very close to close jointed
100–1000	Medium bedded	Medium jointed
Over 1000	Thick bedded	Wide jointed

Basic Geological Mapping, Fifth Edition.
Richard J. Lisle, Peter J. Brabham and John W. Barnes.
© 2011 John Wiley & Sons, Ltd. Published 2011 by John Wiley & Sons, Ltd.

Table D.2 *Abridged grain-size scales.*

Size class (Wentworth)		ϕ-scale	Metric scale
Gravel	Boulders	-8	256 mm
	Cobbles	-6	64 mm
	Pebbles	-2	4 mm
	Granules	-1	2 mm
Sand	Very coarse sand	0	1 mm
	Coarse sand	1	500 µm
	Medium sand	2	250 µm
	Fine sand	3	125 µm
	Very fine sand	4	63 µm
Silt	Silt	–	4 µm
Clay	Clay or mud	–	–

REFERENCES

Badgley, P.C. (1959) *Structural Methods for the Exploration Geologist*, Harper, New York.

Barnes, J.W. (1985) Another home-made clinometer. *Geology Teaching*, **10** (2), 61.

Barnes, J.W. (1988) *Ores and Minerals: Introducing Economic Geology*, John Wiley & Sons Ltd, Chichester.

Barnes, J.W. (1990) The horsetail's tale. *New Scientist*, Feb, issue 1702, p. 74.

Barrass, R. (1978) *Scientists Must Write*, Chapman and Hall, London.

Berkman, D.A. (2001) *Field Geologist's Guide*, Monograph, Vol. 9, 4th edn, Australian Institute of Mining & Metallurgy, Parkville.

Bishop, A.C., Woolley, A.R. and Hamilton, W.R. (2003) *Philip's Minerals Rocks and Fossils*, 2nd edn, Philips, in Association with the Natural History Museum.

Boundary Associates (compilers) (2010) *The Geologist's Directory*, Boundary I-Media, Withyham.

Brabham, P.J. (2009) The Central Valleys of South Wales: Using GIS to visualize the geology, landscape and coal mining legacy, in *Urban Geology in Wales III*, National Museum of Wales Geological Series, Vol. 26 (eds M.G. Bassett, H. Boulton and D. Nichol), Cardiff.

Brooks, R.R. (1983) *Biological Methods of Prospecting for Minerals*, John Wiley & Sons Inc., New York.

Coe, A.L., Argles, T.W., Rothery, D.A. and Spicer, R. (2010) *Geological Field Techniques*, John Wiley & Sons Ltd, Chichester.

Cohee, G.V. (chairman) (1962) *Stratigraphic Nomenclature in the Reports of The United States Geological Survey*, US Government Printing Office, Washington, DC.

Compton, R.R. (1985) *Geology in the Field*, John Wiley & Sons Inc., New York.

Cooper, G.R. and McGillem, C.D. (1967) *Methods of Signal and System Analysis*, Holt, Rinehart and Winston, New York, 432 pp

Crozier, M.J. (1986) *Landslides: Causes and Consequences*, Croom Helm, London.

Basic Geological Mapping, Fifth Edition.
Richard J. Lisle, Peter J. Brabham and John W. Barnes.
© 2011 John Wiley & Sons, Ltd. Published 2011 by John Wiley & Sons, Ltd.

Cruden, D.M. Varnes, D.J. (1996) Landslide types and processes. In: Turner A.K., Shuster R.L. (eds) *Landslides: Investigation and Mitigation*, Transp Res Board, Spec Rep 247, pp. 36–75.

Dietrich, R.V. and Skinner, B.J. (1979) *Rocks and Rock Minerals*, John Wiley & Sons Inc., New York.

Dixon, C. (1999) Letter, *Geoscientist*, **9**, 13.

Duchon, C.E. (1979) Lanczos filtering in one and two dimensions. *J. Applied-Meteor.*, **18**, 1016–1022 .

Evans, A.M. (1993) *Ore Geology and Industrial Minerals: An Introduction*, 3rd edn, Blackwell, Oxford.

Fleuty, M.J. (1964) The description of folds. *Proceedings of the Geologists' Association of London*, **75**, 461–492.

Forrester, J.D. (1946) *Principles of Field and Mining Geology*, John Wiley & Sons Inc., New York.

Fry, N. (1984) *The Field Description of Metamorphic Rocks*, Geological Field Guide Series, John Wiley & Sons Ltd, Chichester.

Geological Society (1972) A concise guide to stratigraphical procedure, *Journal of the Geological Society*, **138**, 295–305.

Geologists' Association (2000) Geological Fieldwork Code, Leaflet, http://www.geolsoc.org.uk/gsl/site/GSL/lang/en/page2542.html.

Gowers, E. (1986) *The Complete Plain Words*, Pelican, London.

Greenly, E. and Williams, H. (1930) *Methods of Geological Surveying*, Thomas Murby, London.

Hansen, W.R. (ed.) (1991) *Suggestions to Authors of the Reports of the United States Geological Survey*, [STA7], 7th edn, U.S. Geological Survey, Reston, VA, p. 289. http://www.nwrc.usgs.gov/lib/lib_sta.htm.

Harbin, P.W. and Bates, R.L. (1984) *Geology of the Non-metallics*, Metal Bulletin Inc., New York.

Heritage, G. and Large, A. (eds) (2009) *Laser Scanner for the Environmental Sciences*, Wiley-Blackwell.

Holland, C.H., Audley-Charles, M.G., Bassett, M.G. *et al.* (1978) *A Guide to Stratigraphical Procedure*, Special Report No. 11, Geological Society, London.

Ingerson, E. (1942) Apparatus for direct measurement of linear structures. *American Mineralogist*, **27**, 721–725.

Johnston, M.R. (1971) Pre-Hawera geology of the Kaka District, North-West Nelson, *New Zealand Journal of Geology and Geophysics*, **14**, 82–108.

Knill, J.L. (ed.) (1978) *Industrial Geology*, Oxford University Press, Oxford.

Lillesand, T.M., Kiefer, R.W. and Chipman, J.W. (2008) *Remote Sensing and Image Interpretation*, 6th edn, John Wiley & Sons Inc., Hoboken.

Liu, G.L. and Mason, P.J. (2009) *Essential Image Processing and GIS for Remote Sensing*, Wiley-Blackwell.

Lisle, R.J. (1980) A simplified workscheme for using block diagrams with the orthographic net. *Journal of Geological Education*, **29**, 81–83.

Lisle, R.J. (2003) *Geological Structures and Maps*, 3rd edn, Butterworth Heinemann, London.

Lisle, R.J. (2006) Google Earth: a new geological teaching resource, *Geology Today*, **22** (1), 33–36.

Lisle, R.J. and Leyshon, P.R. (2004) *Stereographic Projection Techniques in Structural Geology*, 2nd edn, Cambridge University Press, Cambridge.

McCaffrey, K.J.W., Holdsworth, R.E., Clegg, P. *et al.* (2003) Using digital mapping tools and 3D visualisation to improve undergraduate fieldwork, *Planet*, **5**, 34–37.

McCaffrey, K.J.W., Jones, R.R., Holdsworth, R.E. *et al.* (2005) Unlocking the spatial dimension: digital technologies and the future of geoscience fieldwork, *Journal of the Geological Society, London*, **162**, 927–938.

McClay, K.R. (2003) *The Mapping of Geological Structures*, 2nd edn, The Geological Field Guide Series, John Wiley & Sons Ltd, Chichester.

McDonald, C. (2000) Slope stability and landslides, *The Oxford Companion to the Earth*, Oxford University Press, Oxford, pp. 977–980.

Marjoribanks, R. (2010) *Geological Methods in Mineral Exploration and Mining*, 2nd edn, Springer, Berlin.

Marshak, S. and Mitra, G. (1988) *Basic Methods in Structural Geology*, Prentice Hall, New York.

Moseley, F. (1981) *Methods in Field Geology*, Freeman, Oxford/San Francisco.

Opinion (2002) *Geoscientist*, **12** (6), 19.

Pearce, M.A., Jones, R.R., Smith, S.A.F. *et al.* (2006) Numerical analysis of fold curvature using data acquired by high-precision GPS, *Journal of Structural Geology*, **28** (9), 1640–1646.

Perkins, C.R. and Parry, R.B. (1990) *Information Sources in Cartography*, Bowker-Saur, London.

Peters, W.C. (1978) *Exploration and Mining Geology*, John Wiley & Sons Inc., New York.

Prior, D.J., Knipe, R.J., Bates, M.P. *et al.* (1987) Orientation of specimens – essential data for all fields of geology, *Geology*, **15**, 829–831.

Ragan, D.M. (2009) *Structural Geology: An Introduction to Geometrical Techniques*, Cambridge University Press, Cambridge.

Ramsay, J.G. and Huber, M. (1987) *Techniques of Modern Structural Geology*, vol. 2, Academic Press, London.

Ray, R.G. (1969) Aerial Photographs in Geologic Interpretation. USGS Professional Paper 373.

Rowland, S.M., Duebendorfer, E.M. and Schiefelbein, I.M. (2007) *Structural Analysis and Synthesis; A Laboratory Course in Structural Geology*, Blackwell, Oxford.

Scott, P.W. and Bristow, C.M. (eds) (2002) *Industrial Minerals and Extractive Industry Geology*, The Geological Society Publishing House, Bath.

Stow, D.A.V. (2005) *Sedimentary Rocks in the Field: A Colour Guide*, Manson, London.

Thorpe, R. and Brown, G. (1991) *The Field Description of Igneous Rocks*, The Geological Field Guide Series, John Wiley & Sons Ltd, Chichester.

Tucker, M.E. (2003) *Sedimentary Rocks in the Field*, 3rd edn, John Wiley & Sons Ltd, Chichester.

Uren, J. and Price, B. (2010) *Surveying for Engineers*, 5th edn, Palgrave McMillan.

Vansberg, N. (1952) How to write geologese, *Economic Geology*, **47** (2), 220–223.

Vosselman, G. and Maas, H-G. (eds) (2010) *Airborne and Terrestrial Laser Scanning*, CRC Press.

Wallace, S.R. (1975) The Henderson ore body – elements of discovery, reflections, *Mining Engineering*, **27** (6), 34–36.

Wilson, J.P. and Gallant, J.C. (eds) (2000) *Terrain Analysis: Principles and Applications*, John Wiley & Sons Inc.

INDEX

Basic Geological Mapping, Fifth Edition.
Richard J. Lisle, Peter J. Brabham and John W. Barnes.
© 2011 John Wiley & Sons, Ltd. Published 2011 by John Wiley & Sons, Ltd.